Khadim NDIAYE

Sunflower fertilization

Khadim NDIAYE

Sunflower fertilization

Effect of organo-mineral fertilization on sunflower growth and productivity under rainfed conditions

ScienciaScripts

Imprint
Any brand names and product names mentioned in this book are subject to trademark, brand or patent protection and are trademarks or registered trademarks of their respective holders. The use of brand names, product names, common names, trade names, product descriptions etc. even without a particular marking in this work is in no way to be construed to mean that such names may be regarded as unrestricted in respect of trademark and brand protection legislation and could thus be used by anyone.

Cover image: www.ingimage.com

This book is a translation from the original published under ISBN 978-620-6-72529-9.

Publisher:
Sciencia Scripts
is a trademark of
Dodo Books Indian Ocean Ltd. and OmniScriptum S.R.L publishing group

120 High Road, East Finchley, London, N2 9ED, United Kingdom
Str. Armeneasca 28/1, office 1, Chisinau MD-2012, Republic of Moldova, Europe
Printed at: see last page
ISBN: 978-620-8-22693-0

DEDICATION

*Praise be to **Allah**, Lord of creatures, Clement and Almighty, who has granted us health, the chance to enter this prestigious school, ENSA, and the strength and courage to carry out this work. Peace, blessings and salutations be upon the High Prophet **Muhammad (pbuh)**, his family, his companions and his community, as well as to **Sheikh Ahmadou Bamba Khadim Rassoul**, our illustrious, revered guide and namesake.*

***I dedicate this work with deep gratitude** :*

♥ *First of all, to my dear mother, **Mame Diarra Diop**, and my beloved father**, Ngagne Demba Ndiaye**, who have been the pillars of my journey. I pray to the Almighty to keep you with us for a long time to come;*

♥ *To my brothers and sisters: **Mbène (known as Mme Kane), Modou, Sokhna, Awa, Gagnesiry** and*

***Nogoye Ndiaye**, who have always supported and accompanied me at every stage of my life;*

♥ *To my uncle, **Papa Saliou Ndiaye,** whose support has been crucial throughout my academic career, as well as to my dear grandmother**, Satou Ndiaye,** and my aunts, **Ndiollé Diop, Ngoné Diène, Woury Anne, Souna Ndiaye**, as well as to all the members of my family, whether close or distant, for their constant support ;*

♥ *To my twin at ENSA, **Mame Diarra Mbaye** (37ème promotion), for her availability and invaluable involvement in this work;*

♥ ***To all my fellow members of the 38e** ENSA **class**, the elites. I risk mentioning the whole class, just know that you hold a special place in my heart;*

♥ ***To all my godchildren** of the **42ème promotion**, in particular **Diène and Moustapha Thiaré**, **Modou Ngom**, **Salimata Dia**, **Marième Ndiaye**, **Babacar Ibn and Moustapha DIOP**, **Ibrahima Barry**, **Djily Ndiaye**, **Macoumba Ndiaye** ;*

♥ *To my former teachers at primary school, collège, lycée and ENSA, especially the late.*

Pr Talla Gueye, the late Dr Makhourédia Diop, Mr Dame Diokhané, Mr Modou Ndiaye,

***Ibrahima Dieng**, **Thiouba Dieng**, **Seydou Sow** and **Ibrahima Badiane**;*

♥ *To my friends and classmates from primary school, collège, lycée and ENSA **Saliou** and **Pape Gueye**, **Abdou Khadre Djily Beye**, **Khaly Boye**, **Fallou Diop**, **Abdou Fall Dramé**, **Alasane Guissé**, **Oumy Ndiaye**, **Mouhamed Bâ, Malick Ndiaye Diop, Souleymane Sy** and **Cheikh Niang (38ème promotion), Seydou Kâ** and **Ahmed Ndiaye (39ème promotion)**;*

♥ *As well as to all those who in various capacities have given me their help and sympathy.*

ACKNOWLEDGEMENTS

We could not have written this dissertation without their contribution and support, from near or far, in one way or another, at one time or another; we would therefore like to take this opportunity to express our gratitude and heartfelt thanks. In particular, we would like to thank:

Prof. Ibrahima Diédhiou*, Director of ENSA, for the remarkable work he does at the school, both administratively and pedagogically, and for the honour he bestows on me by chairing my jury;*

Dr. Ahmed Tidiane Diallo*, my dissertation supervisor, for his guidance, availability and immense contribution to the preparation of this document;*

Mr Massamba Thiam*, Head of the Plant Production Department, for his teaching, support and advice. Through you, we would like to thank all the other teachers who work in the department;*

Prof. Mamadou Tandiang Diaw*, Director of Studies at ENSA. Through you, we would also like to thank all of ENSA's PER and PATS staff;*

Mr Ibrahima Sarr*, Director of the CNRA in Bambey, who was kind enough to welcome us and who created the necessary conditions for us to achieve our objectives;*

Dr. Malick Ndiaye*, head of the agronomy department at the CNRA in Bambey, for the honour of being our training supervisor. We cannot thank you enough for your availability, your support, your advice and above all your strong involvement in this work. Our warmest thanks go to* ***Youssoupha Sankharé, Modou Ngom, Daouda Faye, Adama Ngom****, and the whole team in the department;*

Mr Aliou Ngom*, manager of the Nioro station, for his kindness, availability and support during our stay. I would also like to make a special mention of* ***Macoumba Jabong and his family, Magueye Anne, Mrs Harandé Ka, Gora Kane, Mamadou Sène, Alasane and his wife Astou, Mané, Mame Samba, Jean****, in short all the staff at the station for their warm welcome and unrivalled hospitality;*

Mrs Oumou Bessane, Mouhamed Tall and his wife Amy, Gora*, and all the members of my host family in Nioro for your kindness and hospitality;*

__My classmates__, the __38ème__, the elite, with whom I shared one of the most beautiful and unforgettable moments of my student life, especially my room-mate from year 1ère to year 5ème, __Modou Saliou SALL__ and my colleagues in the PV 2023 department;

My elders, __Mouhamed Talla Kane (32ème promotion)__, __Adama Sène (34ème promotion)__, __Assane Diop (35ème promotion)__, __Lamine Ngom (36ème promotion)__, __Mame Diarra Mbaye__ and __Aminata Diao (37ème promotion)__ who contributed to the drafting of this document. __Fallou Diouf__, __Daouda Basse__, __Serigne Modou Mbacké__, __Ousmane Pouye__, __Ibrahima Aw__, __Pape Omar Bousso__, __Badara Lô__, __Ahmadou Bamba Ndiaye__, __Khady Ndiaye__, in short the entire __34ème__, __35ème__, __36ème__ and __37ème__ __class of__ ENSA, who supported, accompanied, assisted and advised me throughout my time at the school;

My cadets of the __39ème__, __40ème__, __41ème__ and __42ème__ __promotion__ for the respect and consideration shown to us as well as all the members of the __Dahira Nouroul Mahanhidi of ENSA__, through their __Dieuwrigne Mourtalla Seck.__

SUMMARY

Senegal, a groundnut-producing country, is heavily dependent on imports of vegetable oils, which poses a major challenge to its food security. Against this backdrop, sunflower has been identified as a strategic crop for diversifying local sources of oil and meeting food needs, particularly in the groundnut basin, where the main problems are falling rainfall and land degradation. The development of efficient and appropriate cultivation techniques should therefore be envisaged for sunflower cultivation in Senegal. This study was initiated to assess the effects of organo-mineral fertilisation on the growth and productivity of sunflower (Helianthus annuus L.) under rainfed conditions in the south-central groundnut basin. The trial was conducted at the Nioro du Rip experimental station during the 2023 rainy season. The experimental design was a randomised complete block design with five replications. The factor studied was the fertilisation plan with ten levels: F0 = absolute control; F1=300 kg.ha^{-1} of N-P-K (15-15-15) + 100 kg.ha^{-1} of urea (46-0-0); F2= 150 kg.ha^{-1} of N-P-K (15-15-15) + 50 kg.ha^{-1} of urea (46-0-0); F3=2.5 t.ha^{-1} of "Toss Gui" compost; F4=3.75 t.ha^{-1} of "Toss Gui" compost; F5=5 t.ha^{-1} of "Toss Gui" compost; F6= F1 + F3; F7= F2 + F3; F8= F2+ F4; F9= F2+ F5. The variables monitored and calculated were phenology, growth, yield and its components. The results of the analysis of variance showed that the fertilisation plan did not significantly influence the growth and phenology variables. However, on the yield variables and its components, a significant effect was noted only on seed yield (p=0.0361) and thousand-seed weight (p=0.0196). Overall, treatments F6 and F9 were statistically different from the others. Seed yields were 738.5±199.7 kg.ha^{-1} and 569.5±242.0 kg.ha^{-1} respectively in these two treatments. Head yields were 1504.8±390.3 kg.ha^{-1} (F6) and 1172.9±461.3 kg.ha^{-1} (F9). The biomass recorded was 1140.2±262.8 kg.ha^{-1} for the F6 treatment and 721.9±334.3 kg.ha^{-1} for the F9 treatment. The tallest plants, observed at 60^{e} days after sowing, had an average height of 90.5±5.5 cm (F6) and 87.0±9.1 cm (F9). Flowering and maturity dates were the earliest, at 48±1 and 64±1 days respectively. The average weight of a head and the weight of seeds per head were 51.1±4.6 g and 49.9±6.8 g respectively. Treatments F1 (50,000±18,408 plants.ha^{-1}), F2 (51,250±14,225 plants.ha^{-1}) and F5 (44,750±17,091 plants.ha^{-1}) recorded the highest plant densities. These results show that organo-mineral fertilisation could be recommended to improve sunflower productivity in the south of the Senegalese groundnut basin.

Keywords: Southern groundnut basin, organo-mineral fertilisation, sunflower, seed yield, Senegal.

TABLE OF CONTENTS

INTRODUCTION

Sunflower (Helianthus annuus L.) is the 4^{e} most consumed vegetable oil and the 4^{e} most produced oilseed in the world (USDA, 2018). With almost 29 million hectares under cultivation, annual sunflower production in 2022 is estimated at 54.2 million tonnes of seeds, or 18.5 million tonnes of oil (FAOSTAT, 2023). However, it should be noted that this large production is very unevenly distributed between the continents, with Europe dominating. Africa, for its part, makes a small contribution and relies on imports to make up the shortfall. Sunflower seed production and sunflower oil imports in 2022 are estimated at 2.6 and 1.3 million tonnes respectively (FAOSTAT, 2023). Senegal, one of the world's largest groundnut producers, is unable to produce enough groundnut oil for its inhabitants, who are forced to consume palm, soya or sunflower oil. With a cost of 119.6 million CFA francs spent and a volume of 201,001 tonnes, animal and vegetable oils and fats rank 3^{e} among the most imported consumer goods (ANSD, 2023). Between 2018 and 2021, a total of 61,113 tonnes of sunflower seed oil, or 27% of oil imports in West Africa, were imported (FAOSTAT, 2023). This heavy dependence on imports is undoubtedly a source of food insecurity. With this in mind, the Senegalese government has set up the National Strategy for Food Security and Resilience (SNSAR), which aims to diversify sources of local vegetable oil and protein to help meet the country's food needs. Sunflower, an oil and protein crop, is targeted as a priority crop in the SNSAR. Low in fertiliser consumption, less sensitive to water stress and suitable for all rainfed and irrigated production systems (Agreste, 2014; Yerima et al., 2014), sunflower has a good profile for cultivation in the agro-ecological zone of the Groundnut Basin.With 57% of the country's arable land, the Groundnut Basin remains the country's leading agricultural region (Tounkara et al., 2022). Unfortunately, it is facing a significant drop in rainfall and severe land degradation (Tounkara et al., 2022), resulting from a combination of climatic, soil and human factors. Low soil nutrient availability is a major limiting factor for agricultural production in the region's agrosystems (Tounkara et al., 2020). Farmers also have difficulty accessing these inputs because of their high cost, low income and limited funding. This is why it is necessary to develop fertilisation techniques that are accessible to growers, thereby increasing yields while guaranteeing a sustainable improvement in soil condition in order to meet the demands of climate change, but also to help eradicate famine and food insecurity. Fertilisation Organo-mineral fertilisation, an approach based on the application of a mixture of organic and mineral fertilisers, is emerging as a promising strategy for increasing yields and stabilising them from one year to the next

(Pieri, 1989).In sub-Saharan Africa, much research has demonstrated the importance of combining the use of mineral fertilisers and organic matter in a way that is adapted to local conditions in order to obtain satisfactory cereal yields (Akanza and Yoro, 2003; Akanza et al., 2016; Yerima et al., 2014; Ouandaogo et al., 2016; Somda et al., 2017; Faye, 2022). However, there is a lack of agricultural information on optimal fertilisation and the profitability of sunflower production, particularly in Senegal.

With this in mind, the Senegalese government, through the Institut Sénégalais de Recherches Agricoles (ISRA), decided to collaborate with AGROPOL to carry out on-station fertilisation trials on oilseed crops. The overall aim of this dissertation is to contribute to the promotion of sunflower cultivation in the south of the Senegalese groundnut basin through the development of efficient and appropriate cultivation techniques. More specifically, the aim is to assess the effects of different fertilisation plans on sunflower growth and productivity. The document is structured in three (3) chapters: a literature review on sunflower cultivation and organo-mineral fertilisation, a description of the methodological approach used in this study and a presentation of the results obtained, followed by their discussion. In conclusion, recommendations for future research on the potential impact of this study.
on Senegalese agriculture.

CHAPTER I

BIBLIOGRAPHICAL SUMMARY PART A: SUNFLOWER CULTIVATION

I.1. Taxonomy and classification botany

The scientific name of the sunflower, Helianthus annuus Linnaeus (L.), refers to the characteristic shape of its compound inflorescence, the flower head. It comes from the two Greek words "Helios" and "Anthos", meaning "sun" and "flower" respectively. The sunflower is therefore a kind of "sun flower". The French name comes from the plant's tendency to turn towards the sun during the day (Evon, 2008).

Cultivated sunflower is an annual, diploid plant (2n=34) belonging to the phylum Spermaphytes, subphylum Angiosperms, class Dicotyledons, order Synanthera, family Asteraceae, subfamily Tubuliflora, tribe Heliantheae, genus Helianthus and species annuus (Nouri et al., 2011; Ramde, 2014).

I.2. Origin, domestication and distribution

The domestication of the sunflower began in North America. Archaeological evidence indicates that the Indians cultivated the plant as early as 3,000 BC (before Jesus Christ), probably to extract dyes from the hull and to eat the kernels (Heiser, 1985; Doré and Varoquaux, 2006). The sunflower was then introduced to Europe in the XVIe century, via Spain, where for a long time it was considered only as an ornamental plant. It was the Russians who transformed the sunflower into a field-grown oilseed in the second half of the 18th century^{e} . This is why Russia is considered to be a secondary centre of sunflower domestication. It wasn't until the XIXe century that sunflower cultivation really took off in Eastern Europe, then Western Europe and finally across the Atlantic Ocean to Africa.

I.3. Botanical and physiological characteristics

The main morphological and physiological characteristics, such as height, flower head diameter, length of the vegetative cycle, seed size and oil content, are highly dependent on the soil and climate in which the sunflower is grown (Merrein, 1986). The height of the plant can exceed two metres. In addition to the seed, the plant is made up of three parts (Evon, 2008).

I.3.1. Root system

The main root is of the taproot type and can grow up to 2 m deep (Mazoyer, 2002). Although the taproot can go beyond this depth, it is not very aggressive towards obstacles in the soil (Ramde, 2014). It also develops a large superficial root system. Secondary roots are mainly present on the surface. Their number and diameter decrease with distance from the surface. They are finer and shorter according to Aguirrezabal (1993) and Thebaud (2012) cited by Ramde (2014) **(Figure 1)**.

I.3.2. Aerial part

It comprises the stem (2 to 7 cm in diameter), which is straight, rigid, cylindrical and more or less pubescent depending on the genotype, and the leaves (cordate and more or less toothed, between 20 and 40 per stem in most currently cultivated hybrids (Bonjean, 1993; Nouri et al., 2011)) which are inserted into it (Evon, 2008). The stem of cultivated sunflowers is topped by a single flower head. It tends to bend slightly under the weight of the mature flower head. The degree of curvature of the stem is of fundamental importance, as it determines the angle of the flower head in relation to the stem and therefore the ability to protect the florets and achenes from climatic stress and birds (Bonjean, 1993; Seiler, 1997). Leaves come in a variety of shapes and sizes; the largest are found between the 4^{e} and 10^{e} nodes. They play a major role in the production of the seed's lipid reserves (Evon, 2008). (**Figure 1**).

I.3.3. Chapter

This is the plant's reproductive system. Its diameter can vary on average between 10 and 40 cm in most of the hybrids currently grown. The flower head bears two types of flowers (Nooryazdan, 2009) **(Figure 1)**:peripherally-tied flowers that form one or two rows around the edge of the flower head. There are several dozen of these per plant, and they are purely decorative as they are sterile. They are large and vary in colour from lemon yellow to orange or reddish;

• tubular flowers or florets in the centre, which make up the majority of the flower head. Arranged according to very specific geometric properties, the florets are hermaphroditic. The potential number of florets varies, depending on the diameter of the flower head, from 60 to 3,500.Floral morphology and flowering dynamics mean that sunflowers tend to be allogamous. However, there is considerable variability in the plant's self-incompatibility. Pollination is quasi-entomophilous. Bees and bumblebees are the sunflower's main pollinators

(Demol et al., 2002). The flower is very attractive and the number of visitors is very high.

I.3.4. Seed

According to the description given by Karleskind (1996) and Kartika (2005) cited by Ramde (2014), the sunflower fruit, commonly known as the "seed", is an achene generally consisting of a kernel and a pericarp or shell. The kernel is composed mainly of fat (44%), protein (16%), cellulose (16%), water (7%) and other elements such as minerals and starch (Le Clef and Kemper, 2015; Terres Univia and Terres Inovia, 2022). The fatty acid composition of the seeds can be very rich in linoleic acids (67%) in so-called classic sunflowers, or in oleic acids (80%) in oleic sunflowers (Terres Univia and Terres Inovia, 2022). The hull makes up 20 to 40% of the total weight of the seed (Ramde, 2014) (**Figure 1**).

Fresh buds Harvested seeds

Figure 1: Sunflower morphology (Photos taken by myself)

I.3.5. Vegetative development cycle

The sunflower's complete growth cycle lasts between 100 and 170 days, depending on the variety and growing conditions (OECD, 2006). The sunflower vegetative cycle consists of a succession of five stages: germination and emergence (emergence of the cotyledons), the vegetative phase (establishment of the leaves), flowering (appearance of the flower bud followed by the blossoming of the flower head and formation of the seeds) and ripening (filling of the seeds and drying of the plant) (Evon, 2008). These stages have been described by various authors (Merrien and Milan, 1992; Schneiter and Miller, 1981; Rollier, 1972; Evon, 2008):

• **Germination and emergence (A0-A1)** take 7 to 10 days and depend on seedbed moisture and soil temperature, with a minimum threshold of 4°C and an optimum close to 8°C, according to Terres Inovia (2023) (**Figure 2**);

• **the vegetative phase (A2-B10) or emergence-stage 4-5 pairs of leaves**: this lasts an average of 30 days (Nouri et al., 2011). The plant establishes its aerial parts, in particular the solar energy collectors, as well as its taproot system (**Figure 2**);

• **Flower bud (B11-F1)**: corresponding to the appearance and development of the bud, this is the crop's most active growth phase, with dry matter formation of up to 200 kg. ha^{-1} per day (INA P-G, 2003). This period generally runs from day 40^{e} to day 50^{e} and is characterised by spectacular growth in the leaf surface and root system; it is also the period of maximum absorption of mineral elements (nitrogen, phosphorus, potassium, boron, etc.) (**Figure 2**);

• **Flowering (F1-F3)**: this period, which lasts 15 to 21 days for the plot as a whole and 8 to 10 days per plant, is characterised by protandry, when the anthers release pollen on the first day (male stage), followed by style extension the next day (female stage) (**Figure 2**);

• **the seed filling or priming or maturation phase (M0-M4)**: this is the phase when fatty acids and new proteins are synthesised from amino acids derived from the degradation of leaf and stem proteins. Dry matter growth is limited to around 30 g. ha^{-1} (**Figure 2**).

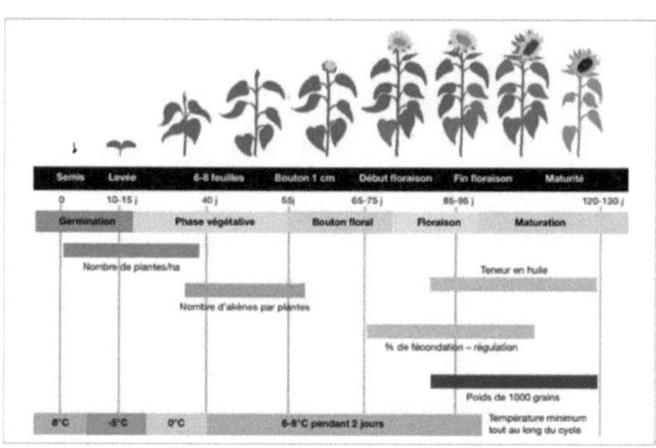

Figure 2: Key stages of sunflower cultivation described by "Mas seeds" in 2024

I.3.6. Heliotropism

The plant's particularity lies in its heliotropism, whereby the sunflower head orients itself to follow the path of the sun throughout the day, from east and south-east in the morning to north-west in the evening, to protect itself from direct sunlight, bad weather and bird attacks

(Evon, 2008). As maturity approaches, the stem bends slightly, both under the weight of the flower head and, above all, as a result of a genetically controlled phenomenon. At the root of this ability is a plant hormone called auxin.

I.4. Ecology

I.4.1. Climate

Sunflowers grow in areas where the average annual temperature varies between 6 and 28°C and where annual rainfall is between 200 and 400 mm. It grows mainly at medium to high altitudes in tropical areas. Young plants can tolerate light frosts. Sunflowers only thrive in areas with a lot of sunshine. However, there is little photoperiodic effect (Vear, 1992).

Thanks to its efficient root system, sunflower is a species that is relatively well adapted to drought, mainly because of its high potential for extracting water from the soil (Hattendorf et al. (1988) cited by Nooryazdan, (2009)). With heat requirements of between 1570 degrees day for early varieties and 1700 degrees day for late varieties (Agridea, 2007), sunflower is less demanding in terms of heat than other summer crops.

I.4.2. Soil

Sunflowers adapt to a wide range of soils, but favour those that warm up quickly. It grows in a wide variety of soil conditions, including laterite, limestone, aluminium toxicity, salinity and sand. Soil pH ranges from 4.5 to 8.7, with an optimum between 6 and 7.2. However, it is sensitive to acidic and waterlogged soils, but not very tolerant of salt, with a drop of around 20% in oil content when salinity exceeds 0.25 to 1.2 $S.m^{-1}$. Germination is only affected if soil salt concentration exceeds 0.7% (Soltner, 2005).

I.5. Agronomy

I.5.1. Soil preparation

Sunflowers are particularly demanding in terms of rooting and emergence density. If the soil is compacted or fragile, deep tillage (20 to 30 cm) with or without ploughing is essential. Direct seeding and very shallow tillage (less than 5 cm) are not recommended for sunflowers in any case, as they do not produce optimum emergence density or quality. sufficient rooting. Any obstacle to pivot development can result in a loss of more than 5 $q. ha^{-1}$ and lower oil content (Terres Inovia, 2023).

I.5.2. Sowing

Sowing requires careful preparation of the seedbed to ensure good soil-seed contact and adequate soil temperature (8 to 10°C in the first 5 cm) for effective, uniform germination and emergence.

The sowing date is determined according to local climatic conditions, so as to reduce the risk of water stress at flowering (Mazoyer, 2002). Late sowing should be avoided as it tends to reduce specific weight (Putnam et al., 1990). [e]Sunflowers are sown around the 2nd half of April or early May (Ahmed, 2022). The Société de Développement et des Fibres Textiles (SODEFITEX) recommends sowing sunflower between 10 and 25 July, following trials carried out in 2004 in the agro-ecological region of Eastern Senegal. If sown before this period, there is a risk of head rot with the October rains. Beyond this period, the cycle may not be completed with an early cessation of the rains (which is very likely). The seed rate is 8 $kg.ha^{-1}$. Seeding can be done with a "Super Eco" animal-drawn seeder. The use of an 18-hole disc gives good emergence regularity along the sowing line (20 cm between bunches) and an acceptable density (66,000 $plants.ha^{-1}$) with a row spacing of 75 cm (SODEFITEX, 2005).

I.5.3. Fertilisation

According to Temagoult (2009), the recommended doses of nitrogen are moderate (no more than 80 units). Sunflowers need 4 to 4.5 kg of absorbed nitrogen per quintal produced (Mas seeds, 2024). When applied in excess, nitrogen favours vegetation exuberance, the development of diseases (Sclerontinia, Phomopsis, Botrytis) and delayed maturity. On the other hand, a nitrogen deficit leads to a loss of yield through a reduction in the number of seeds per flower head and a drop in photosynthetic activity. Nitrogen inputs must therefore be carefully considered. Overfertilising by 50 units means losing 5 $q.ha^{-1}$; underfertilising means losing 4 to 6 $q.ha^{-1}$ (Mas seeds, 2024).

Sunflowers are moderately demanding in terms of potassium and not very demanding in terms of phosphorus. Phosphorus-potassium fertilisation can be limited to 40-60 units per hectare with normal fertilisation, and 50-70 units with reinforced fertilisation (Temagoult, 2009). Exports are 1.2 units of phosphorus and 1.05 units of potassium per quintal produced (Terres Inovia, 2023(a)).

Boron is an essential trace element whose deficiency can lead to a reduction in productivity and oil content. Soils at risk include sandy, chalky soils, compacted with a pH>8 or with low boron levels. It is therefore recommended that preventive boron applications be made to the soil at a rate of 2 $kg.ha^{-1}$ or by foliar fertilisation at doses of 300 to 500 $g.ha^{-1}$, between the 10-leaf stage and the time when the height of the plant is between 55 and 60 cm).

Molybdenum deficiencies in sunflowers are sometimes observed, mainly in plots on acid soils. In general, symptoms are mild and disappear rapidly (Terres Inovia, 2023(a)).

I.5.4. Water requirements and irrigation

Sunflower is a spring crop that is particularly well adapted to drought conditions, thanks to its efficient root system for extracting water from the soil. However, it is sensitive to water stress between the flower bud stage and the end of flowering, requiring an average of 230 mm of water to achieve a yield of over 40 $q.ha^{-1}$. During this critical period, irrigation is particularly beneficial on light soils. Two applications of 35 to 40 mm of water each (the first before flowering and the second after) can increase yield by 8 to 10 $q.ha^{-1}$ and improve oil content by 2 points (Mas seeds, 2024).

I.5.5. Enemies of culture

- **Diseases**

The classification of sunflower pathogens proposed by Guyla et al (1997) makes it possible to distinguish between the various diseases that attack the plant. These include fungal diseases, bacterial leaf diseases and viral diseases (sunflower mosaic virus, sunflower yellow spot, etc.). However, the most serious diseases are caused by fungi, including rust (Puccinia helianthii), downy mildew (Plasmopara halstedii or P. helianthi), verticillium wilt, white (Sclerotinia sclerotium) and grey (Botrytis cinerea) stem and ear rot, phoma (Phoma macdonaldii), phomopsis (Diaporthe helianthi) (Agridea, 2007) and early desiccation syndrome (Bret-Mestries et al., 2016). Sunflowers are also susceptible to other secondary diseases such as alternaria (Alternaria helianthi) and charcoal rot (Macrophomina phaseoli).

Treatments for sunflower fungal diseases are either ineffective or inefficient. The only thing to look for is varietal resistance, hence the great efforts made by breeders (Soltner, 2005). Similarly, cultural practices such as rotation, moderate and balanced fertilisation, good aeration of the soil, seeding at a density that is not too high, good irrigation and sometimes early harvesting are also means of combating the disease. In some cases, fungicide treatments are necessary, either by seed treatment or by spraying at the "tractor limit" stage. (Temagoult, 2009). A four-year rotation is recommended for sunflower, mainly because of its high susceptibility to sclerotinia. In the intervening years, other broadleaf crops should be avoided (Temagoult, 2009).

- **Pests**

Poorly incorporated organic matter, mild temperatures and simplified cultivation techniques that do not destroy or bury crop residues encourage the development of slugs, aphids and wireworms, which cause yellowing of leaves and flower buds or heads, slower growth, irregular emergence and disappearance of seedlings and cotyledons. The presence of slugs at emergence often requires chemical treatment using insecticides such as helicides (Mazoyer, 2002); wireworms are controlled by seed treatment; aphids in the event of an early outbreak at the 2-3 leaf pairs stage (Soltner, 2005).

Bird damage is a major problem in all sunflower-growing regions of the world. It occurs from the start of ripening through to harvest, but seems to be most severe in the first 18 days after anthesis. Small sparrows (Passeridae) and larger species such as crows (Corvidae) and parrots (Psittacidae) eat the seeds (Linz et al., 1997). Acoustic measures (detonations, bird distress calls) are direct and effective means of control, but only temporarily (Agridea, 2007).

- **Weeds**

When it encounters stand problems or when the rotation is not respected, sunflower can develop a specific flora consisting of ragweed, Ammi majus, Xanthium or wild sunflower against which it is not very competitive and will be difficult to control (Lecomte and Nolot, 2011). Control is carried out at sowing time or before emergence, using products with a broad spectrum of action on the most common weeds. Hoeing is a catch-up solution up to the limit of tractor passage (Mazoyer, 2002).

Sunflowers can also be parasitized by broomrape, which causes serious damage in Eastern Europe and Spain (Lepennetier, 2015). Extending rotations by integrating false host species or using varieties with very low to low susceptibility can address the main risks present in the area concerned in the event of moderate to high presence of Orobanche cumana (Terres Inovia, 2023(b)).

I.5.6. Harvest and post-harvest

Sunflowers reach maturity when the underside of the flower head turns yellow and the bracts turn brown, with harvesting periods ranging from late August to early October, depending on various factors. Late harvesting can lead to seed losses due to birds, lodging or disease, while harvesting too early increases impurities and drying costs (INA P-G, 2003). The optimum moisture range for harvesting is between 15% and 9% to achieve maximum yield (Syngenta, 2013).

Harvesting is carried out using a combine fitted with trays with raised edges, fixed in front of the cutterbar to collect the stems and seeds falling in front of the blade (Soltner, 2005). Stem and flower head residues are usually left on the field after the combine has passed through and are later buried with the part of the stem cut during harvesting. This practice returns around 7 tonnes of dry matter per hectare to the soil, equivalent to 1.2 to 1.5 tonnes of humus, which encourages the return to the soil of the mineral elements removed by the plant (Soltner, 1986; Evon, 2008). Threshing involves cutting the sunflower stalks below the flower head, at around half-height. The heads thus freed are threshed in order to detach and recover the seeds. The seeds are then cleaned and dried before being stored in silos and marketed. Before, during or after the combine harvester passes through, seeds may be lost through natural shattering before cutting, in front of the machine or after threshing.

I.6. Importance of growing sunflower

Like soya and rapeseed, sunflower (Helianthus annuus L.) is an annual oilseed crop, grown primarily for the oil content of its seeds and also as a protein crop due to its high protein content (Evon, 2008).

It was cultivated until the XVe century by the American Indians for food purposes (consumption of its seeds raw or in the form of flour) but also for other applications (medicinal, colouring, etc.) (Evon, 2008). Sunflowers were first used for ornamental purposes. It gained popularity for its seeds, which are eaten raw or roasted.

I.6.1. Main applications

Sunflower seed is mainly used for human consumption, either as edible oil (43% of the seed) or as seeds of non-oleaginous varieties. Sunflower meal (55% of the seed) obtained after crushing is considered an alternative source of protein in livestock feed. Finally, sunflower can be used for non-food uses such as the production of biofuel from its oil (Ahmed, 2022; Ramde, 2014; Temagoult, 2009; Ebrahimi, 2008; Mazoyer, 2002; Putnam et al., 1990).

- **Human food**

Sunflower oil is generally considered to be a premium quality oil due to its pale colour, high content of unsaturated fatty acids, absence of linolenic acid and trans fatty acids, neutral flavour, high resistance to oxidation and high smoke points (Ebrahimi, 2008). It is also used in the manufacture of margarine. Sunflower can be eaten toasted or unshelled and is used in processed foods such as cereal bars, bread, etc.

❖ **Animal feed**

Once the oil has been extracted from the sunflower seed, what remains is the meal, which is characterised by its high nitrogen content (45-55%), as well as its richness in methionine and group B vitamins, which are used as a supplementary feed for livestock. Compared with other oilcakes, sunflower oilcake has a better phosphocalcic balance. Sunflower can also be used as a silage crop in areas where the season is too short to produce ripe maize for silage.

❖ **Biofuel**

Faced with expensive and soon to be scarce oil, several countries (mainly developed countries) have turned to biofuels. Sunflower oil can be used either as an agrofuel for diesel engines, or directly as pure vegetable oil (PVO), or as a methyl ester after esterification. With no sulphur emissions, 25% fewer nitrogen by-product emissions and three times less CO2 released during combustion, sunflower fuel is highly relevant today, given the need to reduce greenhouse gases.

❖ **Lipochemistry industry**

The various parts of the sunflower plant can also be used in the manufacture of paints, resins, plastics, soaps, cosmetics, detergents and many other industrial products. The hulls can be used in the production of ethanol and furfural, while the stems are used as a source of fibre for fabrics and paper.

❖ **Use of the plant in beekeeping**

Sunflowers are popular with bees because of their long flowering period and large number of flowers per hectare. Pollen and nectar production varies according to variety, soil and climate. Yields can vary from one to two harvests (16 kg) per hive in a month. Sunflower honey is characterised by its golden colour, fresh aroma and high sweetening power, and is rich in vitamins (Temagoult, 2009).

❖ **Use of the plant in horticulture**

In horticulture, this plant stands out for its spectacular, decorative inflorescences. Depending on the intended use, varieties differ considerably, particularly in terms of branching, number and colour of flower heads, etc. (Temagoult, 2009) (**Figure 3**).

FloristanInca JewelsItalian WhiteMusic Box Sunbeam

Ring of FireSunrich Prado Sunrich LemonSunrich OrangeSundance Kid

Figure 3: Some sunflower varieties used in horticulture (Temagoult, 2009)

I.6.2. Agronomic interest

Sunflower has a number of agronomic and ecological advantages (Sarron, 2016):

- It stands out in particular for its ability to control weed growth, thus facilitating the effective return of subsequent crops;
- its wide range of varieties, which offer appreciable tolerance to various diseases such as phomopsis, sclerotinia and mildew;
- After a sunflower crop, the soil surface is in good structural condition, making it ideal for subsequent planting of other crops such as wheat;
- Early harvesting frees up time for other farming activities, making it an attractive choice for crop rotations;
- from an environmental point of view, sunflowers generally require very little use of plant protection products, which helps to reduce their impact on the environment.

I.7. Achievements and outlook for research

Until 1975, the varieties grown were heterogeneous populations in which it was difficult to combine uniformity and vigour, but since the 1960s, Western Europe has revived cultivation using populations obtained in Russia.

The first F1 hybrid varieties were created in France. This work led to the discovery of nucleocytoplasmic male sterility (Leclercq in 1969). This system, which allows the creation of F1 hybrid varieties, has made it possible to better combine uniformity, productivity, quality and resistance to lodging and disease. Sunflower breeding is marked by a relentless fight against disease. Significant progress has been made in tolerance to sunflower Phomopsis, white head rot and resistance to new races of downy mildew. The significant progress achieved can be explained by the successful use of a number of plant improvement tools: mutagenesis, interspecific crosses, in vitro culture and marking, which complement the selection and field tests that are still the basis of variety creation today. The most important traits improved were :

- yield: a good yield is the main quality required. The switch to F_1 hybrid varieties has made a major contribution to improving yield;
- resistance to lodging: recent varieties are four times more resistant to lodging than varieties developed in the 1970s. It is now quite rare to see sunflower crops lodging. This increased resistance to lodging is an important component of yield stabilisation.

Improving the resistance and agronomic characteristics of sunflowers in drought conditions is now a major environmental and economic challenge. Improving sunflower productivity is based on two complementary approaches: optimising crop management and selecting genotypes with interesting characteristics (Temagoult, 2009).

I.8. Implementation and development of sunflower cultivation in Senegal

Introduced to Senegal in 2004 for experimental purposes, the sunflower extension programme began in 2006/2007 on 58 hectares in two agro-ecological zones (the intermediate zone, represented by the Sare El Hadji site and the Dialakoto multi-local experimentation unit (AMEX), and the northern zone, represented by the Koussanar AMEX and the Aïnouman research and development (R&D) site), with two varieties (ALLIUM and ALL STAR).D) site in Aïnouman), with two varieties (ALLIUM and ALL STAR) for a production of 42,798 kg, i.e. a yield of 738 $kg.ha^{-1}$. In 2007/2008, production marketed was 80,220 kg from an effective area of 150 ha, giving a final yield of 532 $kg.ha^{-1}$. In 2008/2009, production was 45 tonnes on 120 ha and in 2009/2010 20 tonnes on 70 ha, i.e. 300 $kg.ha^{-1}$. Hybrid seed was used for the first 2 seasons, with a subsidy of 75% of the price (CFAF 4,500 per kg). Composite seeds were used for the next two seasons to eliminate the subsidy and reduce the cost of the technical package. Although some individual results are interesting, the majority of growers are struggling to reach 800 $kg.ha^{-1}$, which is a handicap for rapid expansion of the crop. The low yields achieved are attributable to a number of factors, including a lack of mastery of cultivation techniques and the diversion of inputs to other crops deemed more profitable by growers. The sale of the oil is certainly profitable, generating surpluses of over 150 FCFA per litre, but there is a problem with processing the product (SODEFITEX, 2005 and 2011).

In 2023, a multidisciplinary team committed itself to promoting rain-fed sunflower cultivation in Senegal with the aim of fostering food self-sufficiency. The initiative began with a pilot phase covering twenty hectares in the groundnut basin, involving 120 farmers. Building on the success of this initial phase, future plans include buying crops at attractive prices to build up a seed bank, expanding the crop to the Sine-Saloum and Casamance regions by 2024, and processing sunflower into oil and animal feed.

Sunflower has many advantages and its production can be increased and intensified. Private developers (such as AGROPOL) are joining forces with the Senegalese government to increase sunflower production in targeted agro-ecological zones. This initiative is seen as a way of meeting the country's high demand for oil and oilcake and improving the utilisation rate of local agro-industrial units. Sunflower has a number of advantages, including its high oil content (45-50%), its protein-rich meal and the possibility of recycling its residues in agriculture for soil enrichment and energy. It also has the advantage of avoiding the aflatoxin problems associated with groundnuts (Ramde, 2014).

Growing sunflowers in Senegal will help to develop the agro-industry, create wealth and jobs for young people and diversify the sources of oilseeds. However, there is a lack of information and no up-to-date trials on sunflower varieties and fertilisation (the latest studies available to our knowledge date back to 2006). Research is needed to determine which varieties are suited to local conditions (which have changed in the meantime), and the appropriate fertilisation plans and doses for each agro-ecological zone, in order to control sunflower cultivation and supply quality seed to Senegalese growers.

PART B: ORGANO- MINERAL FERTILISATION

I.1. Definition

Falisse and Lambert (1994) define fertilisation as a set of coordinated cultivation practices aimed at ensuring that cultivated plants are correctly supplied with nutrients, through the addition of fertilising materials such as fertilisers and soil improvers.

Fertilisers are substances, usually mixtures of mineral elements, designed to provide plants with additional nutrients to improve their growth and increase crop yields and product quality. Soil conditioners are substances used to improve the physical, chemical and biological characteristics of the soil. They are also classified into two types: mineral amendments, such as limestone and magnesium, and organic amendments, which include materials such as manure, compost, crop debris, slurry and "green manures" (Ndiaye and Hereau, 2012).

I.2. Typology of fertilisation

I.2.1. Organic fertilisation

Organic fertilisation is based on the addition of elements such as compost, crop residues, plant waste, animal droppings and organic fertilisers (Ouédraogo et al., 2001; Diagne, 2004; Faye, 2022; Siboukeur, 2013). Organic resources improve water retention capacity, reduce erosion, increase infiltration and improve the soil cover rate through their impact on cation exchange capacity (CEC), root penetration and soil organisms. The contribution of organic resources enables the nutrients provided in the form of fertilisers to be used efficiently (Abga, 2013).

I.2.2. Mineral fertilisation

The aim of mineral fertilisation is to provide the plant with all the mineral elements it needs in sufficient quantities, when it needs them. This is achieved by applying mineral fertilisers, which can be divided into 2 groups (Soltner, 2011): soluble mineral fertilisers and insoluble or poorly soluble mineral fertilisers. Soluble mineral fertilisers are made up of :

• straight fertilisers, which contain only one major nutrient. This category includes nitrogen fertilisers (nitric, ammoniacal and ureic), phosphate fertilisers produced by acid attack (superphosphates, ammonia phosphate) and potassium fertilisers.

• compound fertilisers, which can be binary (NP, NK, PK) or ternary (NPK).

I.2.3. Organo-mineral fertilisation

The combination of organic and mineral fertilisers creates the ideal environmental conditions for growing crops. Organic fertilisers improve the properties of the soil, while mineral fertilisers provide plants with the necessary nutrients. Organic fertiliser is not enough to ensure the agricultural production we need. It must therefore be supplemented by mineral fertilisers (Fertial, 2017).

Organic fertilisation combined with mineral fertilisers solves the problems of falling soil organic matter content and nutrition, while maintaining high and stable crop yields (Dembélé, 1994; Zeinabou et al., 2014; Akanza et al., 2016; Somda et al., 2017 cited by Ndiaye et al., 2019). Organo-mineral and organic fertilisation methods appear to be better at preserving production potential and cause less soil degradation (N'Tarla, 1989; Dembélé, 1994 cited by Ndiaye et al., 2019). According to Abga (2013), the combination of organic and mineral inputs prevents nitrogen immobilisation.

According to Tounkara et al (2022), the benefits of combining mineral fertiliser and organic matter are crucial in a context where farmers in the groundnut basin have long been using organic fertilisation practices that vary greatly from plot to plot. These same authors went on to cite various researchers who have demonstrated the positive impact of organo-mineral fertilisation on the yield of certain species. Pieri (1989) states that organic fertilisation combined with mineral fertilisation increases yields and stabilises them from one year to the next. According to Berger (1996), a mineral supplement in the form of fertiliser is generally necessary to meet the crop's immediate needs. The positive effect of joint application of mineral and organic fertilisers has been noted for maize (Nyami et al., 2014), millet (Badiane et al., 2001; Coly et al., 2021) and sorghum (Somda et al., 2017). Furthermore, Tittonell and Giller (2013) report that the absence of sufficient organic restitution on a long-term continuous crop could degrade soils to the point of making them 'non-sensitive' to mineral fertilisers, hence the combination of the two manures (organic and mineral) to protect soils and improve their productivity.

CHAPTER II

MATERIALS AND METHODS

II.1. Presentation of the study area

The Nioro station site is located between 13°45' north latitude and 15°46' west longitude. The study area is in the southern part of the agro-ecological zone of the Groundnut Basin and belongs to the Kaolack region and the department of Nioro du Rip. It is strategically located at the north-west entrance to the commune of Nioro du Rip (**Figure 4**). From an agronomic point of view, this location offers a good representation of the southern zones of the groundnut basin, close to the 700 mm isohyet according to Nsome (1999). The climate in this area is of the Sudano-Sahelian type, hot and dry (BWh). It is characterised by the alternation of two main seasons:

- a hot and humid season, from June to October, corresponding to the rainy season;
- a long dry season, lasting around 7 to 8 months, interrupted by a cool period lasting 2 to 3 months (from November to January).

The station is at the heart of the Baobolong valley, a tributary of the Gambia River, located on the right bank (Beye et al., 2012). The station's soils, generally positioned on a slope, are exposed to water erosion due to rainwater runoff. They are characterised by a reddish colour and low humus content (Nsome, 1999).

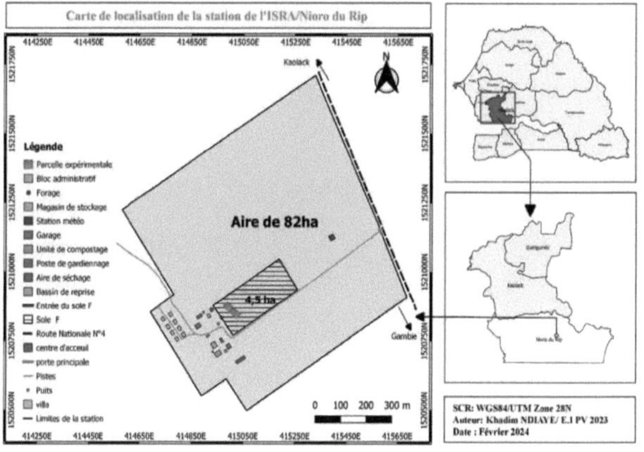

Figure 4: Map showing the geographical location of the study area

The meteorological measurements taken at the Nioro station included rainfall, temperature and relative humidity (minimum and maximum). These data were collected daily, from sowing to harvesting. The rainy season lasted about five months (from June to October). The first rainfall was recorded on 14 June 2023 (1.9 mm). Cumulative rainfall for the 2023 rainy season was estimated at 858.5 mm spread over 46 days, with July, August and September being the wettest months. The heaviest rainfall was recorded the day before sowing, on 24 August, with an intensity of 133.3 mm. The second dekad of October marked the end of the 2023 rainy season, according to the national civil aviation and meteorology agency (ANACIM) (2023) (**Figure 5**). Throughout the trial, maximum temperatures ranged from 29.9 to 41.3°C, and minimum temperatures from 12.2 to 26.4°C, with an average of 29.4°C. Relative humidity showed significant fluctuation, ranging from 11% to 79% for the minima, and from 58% to 97% for the maxima (**Figure 6**).

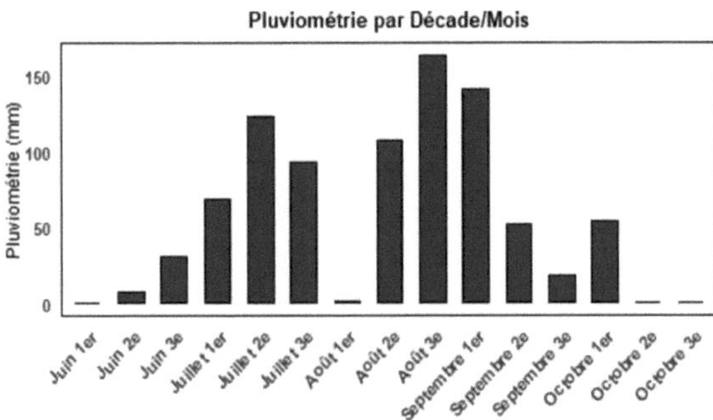

Figure 5: Cumulative rainfall over ten days at the Nioro du Rip station (ANACIM, 2023)

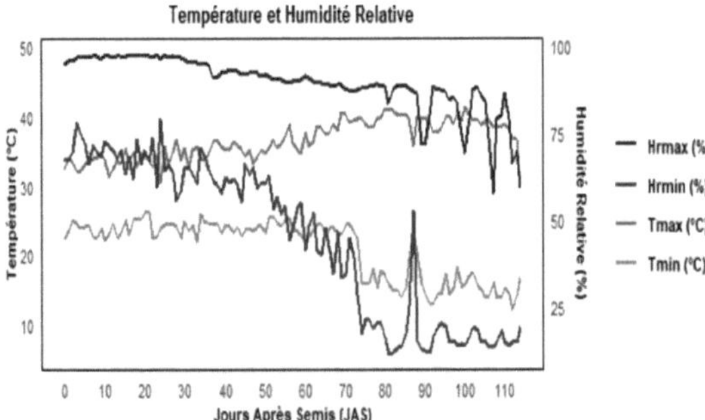

Figure 6: Average ten-day temperature and relative humidity at the Nioro du Rip station (ANACIM, 2023) after sowing date.

Before the trial began, three composite samples were taken at depths of 0-20 cm and 20-40 cm. Each sample was then analysed to determine particle size, pH (water and KCl), total nitrogen, total carbon, C/N ratio, assimilable phosphorus, bases, etc. (Ca^{2+} , Mg^{2+} , Na^{+} , K^{+}), cation exchange capacity and organic matter at the CNRA soil-water-plant laboratory in Bambey. Overall, the soil at the trial site has a sandy texture and is slightly acidic. The organic matter content was relatively low in both horizons. Total nitrogen, total carbon and exchangeable base contents vary from low to moderate, while assimilable phosphorus is moderate in the 0-20 cm horizon and slightly high in the 20-40 cm horizon. Cation exchange capacity is low in both horizons (**Table 1**).

Table 1: Results of soil analysis before sowing (CNRA, 2023)

Horizon.	Granulométrie		
	Arg.	Li.	Sa.
0-20cm	4,74	4,62	90,64
20-40cm	11,45	4,01	84,54

Horizon	Paramètres chimiques											
	pH	CE	NT	PAss	K^{+}	Ca^{2+}	Mg^{2+}	Na^{+}	CEC	MO	C/T	C/N
0-20cm	6,58	30,53	0,030	9,592	0,075	0,723	0,028	0,084	1,915	0,48	0,279	9,24
20-40cm	5,94	22,42	0,023	12,047	0,074	1,789	0,698	0,078	2,926	0,31	0,182	7,95

EC=Electrical Conductivity in µS/cm; NT=Total **Nitrogen** in %; pH=Hydrogen **Potential** (ratio 1/2.5); APss=Assimilable **Phosphorus** in ppm; CEC=Cation Exchange **Capacity** in meq/100g; C/T=Total **Carbon** in %; MO=Organic **Matter** in %; **ppm=parts** per million; **µS=micro** Siemens; **meq/100g** = milliequivalent per 100g of soil; $\mathbf{K^{+}}$ =potassium in meq/100g; $\mathbf{Ca^{2+}}$ =calcium in meq/100g; $\mathbf{Mg^{2+}}$ =Magnesium in meq/100g; $\mathbf{Na^{+}}$ =sodium in meq/100g; **Arg.**=clay in %; **Li.**=limestone in %; **Sa.**=sand in %.

II.2.Material plants

In this study, the hydrid variety **Es Veronika**, registered in the European Union (EU) in 2016, was used as plant material. It is a linoleic, semi-early variety. It is characterised by very good initial vigour, high plant height and a semi-turned head. Not very susceptible to lodging, **Veronika** has an average thousand-seed weight (44.7 to 50.4 g), flowers mid-late and harvests mid-early. Its very high oil content and excellent yields make it a high-performance variety. In terms of disease resistance, it is classified by Terres Inovia and Lidea as tolerant to Phomopsis and Verticillium, moderately susceptible to Sclerotinia head blight and not very susceptible to Sclerotinia crown blight. **Veronika** adapts well to a variety of soil and climatic conditions, withstanding moderate stress in hot, dry environments, as well as cold, wet conditions. To optimise emergence, we recommend aiming for 55-70000 plants.ha^{-1} , depending on the potential of the plot. For limiting potential, aim for 55-60000 plants.ha^{-1} , for good potential, 60-65000 plants.ha^{-1} , and for high potential, 65-70000 plants.ha^{-1} . These characteristics and growing recommendations make **Veronika** a variety that is well suited to different agricultural situations (Terre Inovia and Lidea, 2023).

II.3. Mineral fertilisers and organic soil improvers used

For mineral fertilisation, two types of mineral fertiliser were used:

- **NPK triple 15** i.e. **15N-15P-15K** ;
- **granulated urea 46N-0P-0K**.

Toss Gui" compost, an organic soil improver produced by the controlled fermentation of specific micro-organisms on organic matter selected for its humus content, was used as an organic fertiliser. It is a powdery product (particle size < 10 mm) packaged in 50 kg laminated bags, manufactured and marketed in Senegal by the company Sahélienne d'Entreprise de Distribution et d'Agro-business (SEDAB) (**Table 2**). It can be used for all types of soil and all types of crop, starting at 30g.m^2 (300 to 800 kg.ha^{-1}) depending on the fertilisation plan, and up to 1,500 kg.ha^{-1} when soils are low in organic matter (according to SEDAB/BIOTOSS).

Table 2: Composition of "Toss Gui" compost (according to SEDAB/BIOTOSS)

Eléments	N (%)	P (%)	K (%)	Humidité (%)	MO (%)	pH	C/N	Oligo-éléments
Valeurs	3	2	2	30	> 65	7	16 à 22	Mn, Fe, Mg, Cu, B, Zn, Mo

II.4. Experimental set-up and factor studied

The experimental design used was a randomised complete block design with five replications (**Figure 7**). The factor studied was the fertilisation plan with 10 levels (**Table 3**). In addition to the control (F0), the mineral fertilisation consisted of two modalities (F1 and F2), the amendment included three variants (F3, F4, F5) and the organo-mineral fertilisation consisted of four modalities (F6, F7, F8, F9). The different treatments were obtained by combining four doses of compost (0 t. ha^{-1} , 2.5 t. ha^{-1} , 3.75 t. ha^{-1} and 5 t. ha^{-1}) with three doses of mineral fertiliser (0%, 50% and 100%) (**Table 3**). Each experimental unit was represented by an elementary plot in the form of a square. In this configuration, eight (8) 4 m rows were dedicated to each plot. Thus, each elementary plot had a surface area of 16 m² and 208 bunches. However, only the six (6) The central plots, covering an area of 12 m², were considered useful for data collection. The distance between blocks was 1.5 m and 1 m between plots within the same block (**Figure** 7).

Table 3: Characteristics of the fertilisation plan tested on sunflower

	Compost	Bottom dressing	Cover fertilizer	
Treatments	"Toss Gui (t. ha)-1	NPK (15-15-15) (kg. ha)-1	Urea 46%N (kg. ha)-1	Combinations
F0	0	0	0	0
F1	0	300	100	300 kg.ha-1 + 100 kg.ha-1
F2	0	150	50	150 kg.ha-1 + 50 kg.ha-1
F3	2,5	0	0	2.5 t.ha-1
F4	3,75	0	0	3.75 t.ha-1
F5	5	0	0	5 t.ha-1
F6	2,5	300	100	F3 + F1
F7	2,5	150	50	F3 + F2
F8	3,75	150	50	F4 + F2
F9	5	150	50	F5 + F2

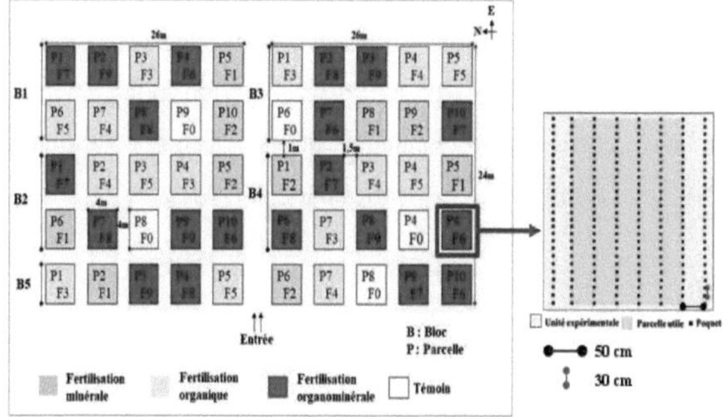

Figure 7: Diagram of the experimental set-up

II.5. Conducting the trial

The land on which the trial took place had been lying fallow for over a year. Soil preparation began with deep ploughing using a tractor, followed by harrowing to ensure an optimum seedbed and thus create favourable soil conditions for good germination and seedling emergence. The seed was sown on 25 August at the Nioro du Rip experimental station during the following period wintering 2023 after a rainfall of 133.3 mm. Two seeds were sown in each poquet, with a spacing of 50 cm between rows and 30 cm between poquets. Due to a low emergence rate, a second sowing (reseeding) was carried out two weeks later, on 9 September 2023. Ten days after this reseeding, a hoeing operation was carried out, followed by the removal of one plant per cluster and the transplanting of the missing clusters. As for the fertilisation plan, two applications were made: the first, including the first dose of urea, was applied 20 days after reseeding; the second, corresponding to the second dose of urea, was applied 22 days after the first (**Table 4**). The compost and triple 15 were broadcast, while the urea was ploughed into the soil at a depth of 5 to 10 cm and at a distance of a few centimetres from the plants, followed by light scratching to encourage rapid incorporation. A second hoeing was carried out after the fertiliser was applied. To compensate for breaks in rainfall, supplementary sprinkler irrigation was applied twice a week between the second dekad of September and the first dekad of October, with the aim of irrigating up to the field capacity of the soil. Harvesting was carried out manually on 107^{e} days after sowing. In each elementary plot, eight lines were considered as yield squares instead of the six initially planned, due to the gaps observed in the majority of plots. Samples of harvested heads and biomass (stems + leaves) were weighed, then reweighed after a two-week period in the drying area at the Nioro station. Threshing was carried out manually at the Centre National de Recherches Agronomiques (CNRA) in Bambey, 46 days after harvest (JAR).

Table 4: Fertilisation plan applied to the individual plot

Doses applied per elementary plot fertilization

First contribution **(29/09/2023)**

Second contribution **(21/10/2023)**

F1	480g N-P-K (15-15-15) + 80g urea (46-0-0)	80g urea (46-0-0)
F2	240g N-P-K (15-15-15) + 40g urea (46-0-0)	40g urea (46-0-0)
F3	4 kg of "Toss Gui" compost	-
F4	6 kg of "Toss Gui" compost	-
F5	8 kg of "Toss Gui" compost	-
F6	F3 + F1	80 g urea (46-0-0)
F7	F3 + F2	40g urea (46-0-0)
F8	F4 + F2	40g urea (46-0-0)
F9	F5 + F2	40g urea (46-0-0)

II.6. Measurements and observations

To assess the effect of organo-mineral fertilisation on sunflower growth and yield, a number of measurements and observations were made on edapho-climatic and agromorphological variables, as well as on yield parameters and their components.

Agromorphological parameters

The agromorphological variables studied include :

• **Growth and development variables** :

○ **Plant density**: this corresponds to the number of plants per unit area after germination and plant establishment. It was taken at 15^{e} , 30^{e} and 60^{e} days after sowing (DAS). It is expressed as the number of plants.ha^{-1} and is calculated using the formula :

$$Dens = \frac{\text{NbPl}}{\text{sup}}$$

where **NbPl** represents the number of plants raised per plot, and **sup**, the area of the elementary plot.

○ **Plant height**: This measurement was carried out in all the experimental units on 10 randomly selected patches within each elementary plot, at 30^{e} and 60^{e} JAS. Height was measured from the base of the stem (collar) to the apical end. Mean values, expressed in centimetres (cm), were then calculated for each plot.

• **Phenological variables**

o **Duration of the sowing-50% flowering cycle**: This is the number of days between sowing and the opening of 50% of the flowers in each plot. To determine this variable, the sowing and 50% flowering dates were recorded in each plot.

DSF = $Date\ de\ 50\%\ floraison$ - $Date\ de\ semis$

o **Duration of the sowing-50% maturity cycle**: This represents the number of days between sowing and the maturity of 50% of the plants in each plot. To determine this variable, the dates of sowing and 50% maturity were recorded in each plot.

DSM = $Date\ de\ 50\%\ maturité$ - $Date\ de\ semis$

Performance variables and performance components

For each experimental unit, the yield and its components were determined at harvest on the basis of the yield square selected in the useful plot. Measurements included seed, biomass and flower head yields, average weight of a flower head, number of flower heads per plant, the number of seeds per flower head, the weight of seeds per flower head and the weight of a thousand seeds.

➢ **Determination of performance variables.**

Before calculating the yield parameters, the following data were collected at the time of harvest and after drying:

o Number of bins harvested per elementary plot = NbPoR ;

o Total number of flower heads (standing or fallen) harvested per elementary plot = NbCapR;

o Useful harvested area (in square metres) =SuR=Spacing x NbPoR =(0.5 x0.3) x NbPoR ;

o Dry weight of flower heads per elementary plot = PCapS. (in grams) ;

o Weight of dry above-ground biomass (stems + leaves) elementary plot= PBms (in grams) ;

o Dry weight of seeds per elementary plot =PGrS. (in grams).

All the values of the yield variables obtained after the calculations are estimated in $kg.ha^{-1}$ (**Table 5**).

Table 5: Yield variables

Variables	Abréviations	Formules de calcul	Unités
Rendement en graines	RdtGr.	$\frac{PGrS}{SuR}$	
Rendement en capitules	RdtCap.	$\frac{PCap}{SuR}$	$kg.ha^{-1}$
Rendement en biomasse	RdtBiom	$\frac{PBms}{SuR}$	

➢ **Determination of yield components (Table 6).**

Table 6: Yield component variables

Variables	Abréviations	Formules de calcul	Unités
Nombre de capitules par plante	NbCap/pl.	$\frac{NbCapR}{NbPoR}$	
Poids des graines par capitule	PGr/Cap.	$\frac{PGrS}{NbCapR}$	g
Poids de mille graines	PMG	$\frac{A \times 1000)}{N}$	g
Poids moyen d'un capitule	Pmoy.Cap	$\frac{PCapS}{NbCapR}$	g
Nombre de graines par capitule	NbGr/Cap	$\frac{(1000 \times PGr.Cap)}{PMG}$	

A= Poids de l'échantillon après séchage à l'étuve ; N= Nombre de graines de l'échantillon

II.7. Analysis of data

The data collected was first entered into an EXCEL spreadsheet and then subjected to an analysis of variance (ANOVA) at the 5% threshold, after checking the validity conditions (independence, normality and homogeneity). The Tukey test was used to compare the means at the 5% probability threshold in cases where significant effects were detected in the ANOVA analysis. A principal component analysis (PCA) was also performed on the standardised data matrix to better visualise the links between the traits measured and the different types of fertilisation studied. Finally, a hierarchical ascending classification (HAC) was carried out to distinguish the different groups and their behaviour in relation to the variables studied.All analyses were performed using R software version 4.2.2.

CHAPTER III

RESULTS AND DISCUSSION

III.1. RESULTS

III.1.1. Effect of fertilisation on growth and development parameters

➢ **The height of the plant**

Analyses of variance provided p-values of 0.178 (at 30^{e} days) and 0.167 (at 60^{e} days) at the 5% threshold, showing that there was no statistically significant difference between plant heights whatever the fertilisation plan applied (**Table 7**). Average sunflower plant heights ranged from 29.8 ± 4.4 cm (F0) to 38.3 ± 3.8 cm (F6) at 30^{e} days after sowing (DAS), with an average of 33.5 ± 4.5 cm. At 60^{e} days after sowing, the average rose from 33.5 to 83.9 ±7.9 cm. However, arithmetically, the tallest plants were obtained with treatments F6 (respectively 38.3±3.8 cm and 90.5±5.5 cm at 30^{e} and 60^{e}), F9 (respectively 34,3±5.8 cm and 87.0±9.1 cm respectively at 30^{e} and 60^{e}) and F4 (35.3±4.1 cm and 84.9±7.5 cm respectively at 30^{e} and 60^{e}) and the shortest with treatments F0 and F2 (**Figure 8**).

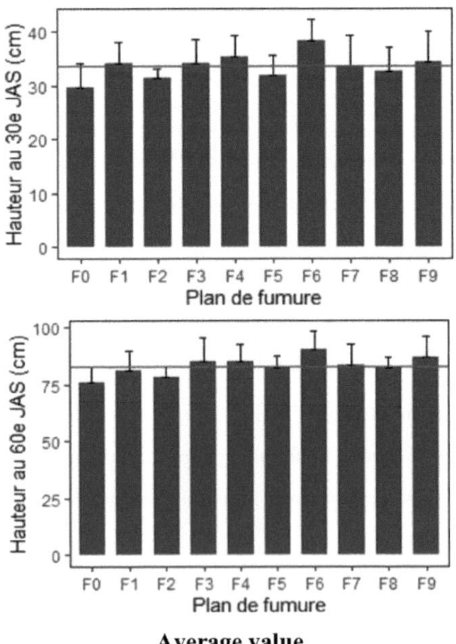

Average value

Figure 8: Average plant height at 30^{e} and 60^{e} days after sowing according to the fertilisation plan applied.

Table 7: Effect of fertilisation plan on plant height

Statistical parameters	Height des plants (cm)	
	30e JAS	60e JAS
Mean±Standard deviation	33,5±4,5	83,9±7,9
Min	29,8	78,1
Max	38,3	90,5
CV	12,75%	9,01%
Pr > (F)	0,178ns	0,167 ns

DAS =Days after sowing; CV=Coefficient of variation; ns=not significant at the 5% level; **Pr=probability** at the 5% level; **cm=centimetre**; **Min=minimum**; **Max=maximum**.

➢ **Plant density**

Analysis of variance did not show any significant differences (**Table 8**) between the different treatments. Average plant density was estimated at 17,350±5,913 plants.ha^{-1} , 42,500±16,339 plants.ha^{-1} and 41,150±15,485 plants.ha^{-1} at 15^{e} , 30^{e} and 60^{e} JAS respectively. Overall, the lowest plant densities were recorded at 15^{e} JAS in the plots receiving treatments F0 (15,000±3,030 plants.ha^{-1}) and F9 (13,250±5,027 plants.ha^{-1}), whereas were obtained with F1(50,000±18,408 plants.ha^{-1}) and F2 (51,250±14,225 plants.ha^{-1}) at 30^{e} JAS (**Figure 9**).

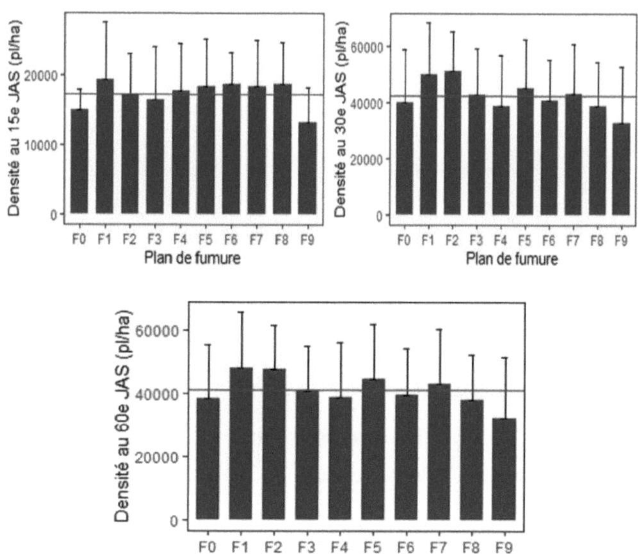

Figure 9: Average plant density at 15^{e} , 30^{e} and 60^{e} days after sowing according to the fertilisation plan applied.

Table 8: Effect of fertilisation plan on plant density

Statistical parameters Plant density (plants.ha)$^{-1}$

	15e JAS	30e JAS	60e JAS
Mean±Standard deviation	17 350±5 913	42 500±16 339	41 150±15 485
Min	13 250	33 000	48 000
Max	19 375	51 250	32 375
CV	30,34%	23,73%	23,46%
Pr > (F)	0,729ns	0,21ns	0,297ns

DAS =Days after sowing; CV=Coefficient of variation; ns=not significant at the 5% level; **Pr=probability** at the 5% level; **cm=centimetre; ha=hectare; Min=minimum; Max=maximum.**

➢ **The length of the cycle sowing-50% flowering and sowing-50% maturity**

At the 5% threshold, analysis of variance showed that the phenological parameters were not significantly influenced by the fertilisation plan (**Table 9**). The average duration of the cycle from sowing to 50% flowering (DSF) and from sowing to 50% maturity (DSM) was 48±1 days and 64±1 days respectively. However, arithmetically, the earliest plants began to flower from 46^{e} days (F0 and F9) and reached maturity two weeks after flowering, i.e. 63 days (except F1 and F4). The latest durations were obtained with treatments F1 and F4 (50 days for flowering at 50% and 65 for maturity at 50%) (**Figure 10**).

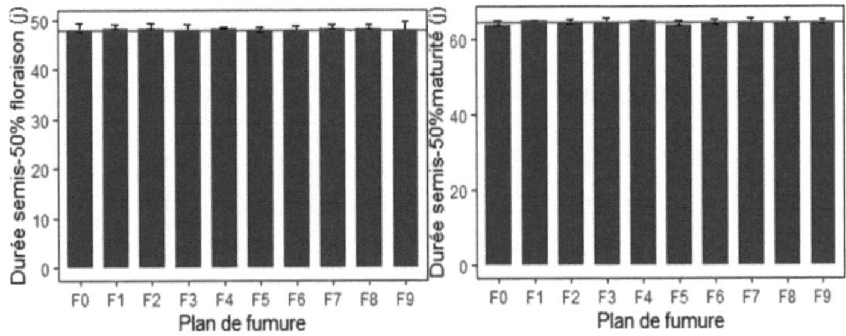

Average value

Figure 10: Duration of sowing-50% flowering and sowing-50% maturity as a function of the fertilisation plan

Table 9: Effect of fertilisation plan on phenology

Variables	Mean± Standard deviation	Min	Max	CV	Pr > (F)
DFS	48±1	46	50	1,39%	0,969 ns
DSM	64±1	63	65	1,35%	0,305ns

DSF=duration of cycle from sowing to 50% flowering (in days); DSM=duration of cycle from sowing to 50% maturity (in days) **CV=coefficient** of variation; **Pr=probability** at 5% level; **ns=not** significant at 5% level; **Min=minimum**; **Max=maximum**.

III.1.2. Effect of fertilisation on yield and its components

➢ **Performance**

The results of the analysis of variance showed that the fertilisation plan had a significant effect on seed yield at the 5% threshold (**Table 10**). The average seed yield was estimated at 498.9±207.6 $kg.ha^{-1}$. Statistically, the plots receiving treatments F6, F5, F9 and F4 were the most productive, with yields of 738.5±199.7^{a} $kg.ha^{-1}$, 570.2±148.3^{ab} $kg.ha^{-1}$, 569.5±242.0^{ab} $kg.ha^{-1}$ and 561.1±157.6^{ab} $kg.ha^{-1}$. Conversely, the control treatments F0 (273.5±121.2^{b} $kg.ha^{-1}$) and F2 (348.8±102.7^{ab} $kg.ha^{-1}$) recorded the lowest yields (**Figure 11**). However, biomass and flower head yields did not show any significant differences in the ANOVA. The average biomass yield in the different treatments was estimated at 718.2±385.7 $kg.ha^{-1}$. This varied from 453.7 $kg.ha^{-1}$ (F0) to 1,140.2 $kg.ha^{-1}$ (F6). With an average of 1,065.1±412.9 $kg.ha^{-1}$, flower head yield exceeded 1 $tonne.ha^{-1}$ in all treatments except F0, F1, F2, and F3. Treatments F6, F5, and F9 reached record values of 1,504.8±390.39 $kg.ha^{-1}$, 1,187.7±331.7 $kg.ha^{-1}$, and 1,172.9±461.3 $kg.ha^{-1}$ respectively (**Figure 11**). Overall, yields were significantly higher in plots where organic or organo-mineral fertilisation was used, compared with the control and mineral fertilisation.

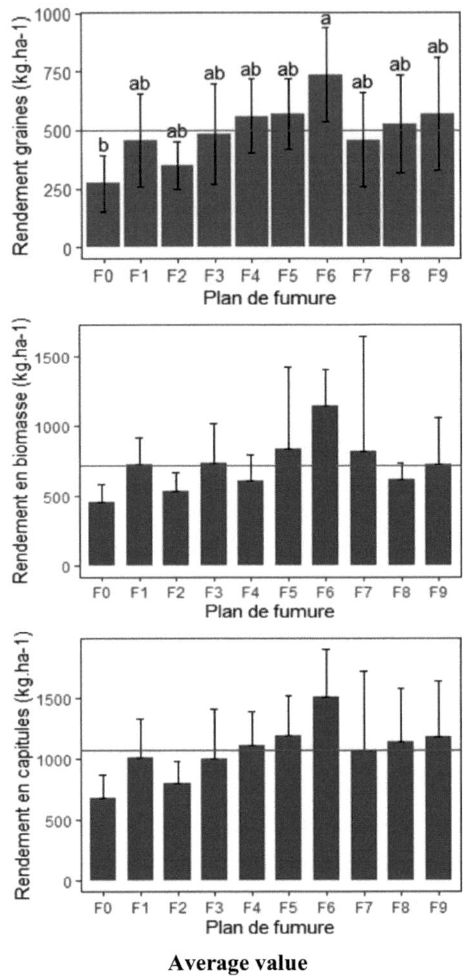

Average value

Figure 11: Average seed, biomass and flower head yields according to fertilisation plan

Table 10: Effect of fertilisation plan on yield

Statistical parameter			Yield (kg.ha-1)	
Seeds			Biomass	Flower heads
Mean±Standard deviation	498,9±207,6	718,2±385,7	1065,1±412,9	
Min		273,5	453,7	673,5
Max		738,5	1140,2	1504,8
CV		37,90%	51,04%	37,50%
Pr > (F)		0,0361*	0.244^{ns}	0.155^{ns}

Pr=probability at the 5% level; CV=coefficient of variation; ns=not significant at the 5% level; *=significance at the 5% threshold; Min=minimum; **Max=maximum.**

- **Performance components**

Analysis of variance at the 5% threshold showed a non-significant effect of the fertilisation plan on the weight of a flower head (Pmoy.Cap), the weight of seeds per flower head (PGr/Cap) and the number of seeds per flower head (NbGr/Cap) between treatments (**Table 11**). For the average weight of a flower head, treatments F9 (18.39±6.26 g) and F6 (17.94±4.56 g) recorded the highest weights. Conversely, F2 (10.87±2.10 g) and F3 (12.61±3.06 g) gave the lowest values. The average value for all treatments was estimated at 15.1±5.1 g. In terms of seed weight per flower head, F9 (8.91±3.25 g) and F6 (8.75±2.05 g) gave the highest values. Conversely, F2 (4.75±1.30 g) and F1 (6.05±1.72 g) had the lowest weights. The average value for all treatments was 7.2±2.4 g. In terms of the number of seeds per flower head, treatments F0 (178±31 seeds) and F9 (175±40 seeds) gave the highest weights. The number of seeds per flower head was the highest, with values above the average (157±34 seeds). Treatment F3 recorded the lowest number of seeds per flower head with an average of 140±27 seeds (**Figure 12**). On the other hand, the thousand-seed weight (TKW) was significantly affected (**Table 11**) by the fertilisation plan. Statistically, the thousand-seed weight varied from $38.4±4.4^a$ g to $51.1±4.6^b$ g for the different treatments, with an average of 44.3±6.9 g. The average thousandweight was estimated at 44.3±6.9 g Arithmetically, treatments F6, F9, F8 and F4 were the only ones to record above-average values. The F0 control obtained the lowest thousand-seed weight, followed by treatments F1, F2, F5 and F7 (**Figure 12**).

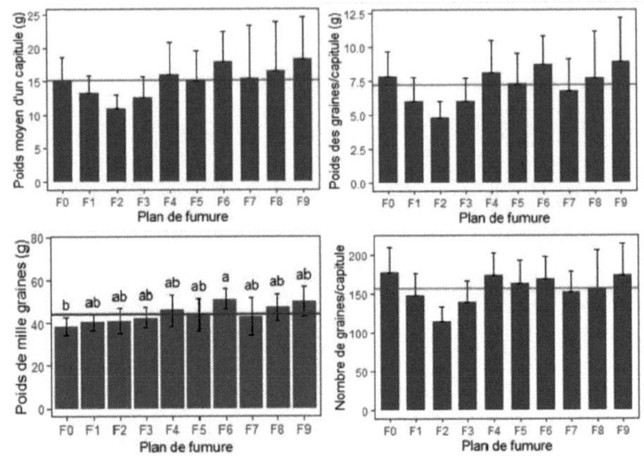

Figure 12: Yield components as a function of the fertilisation plan

Table 11: Effect of fertilisation plan on yield components

Statistical parameterPerformance components

	PMG (g)	Pmoy.Cap (g)	PGr/Cap (g)	NbGr/Cap
Mean±Standard deviation	44,3±6,9	15,1±5,1	7,2±2,4	157±34
Min	38,4	10,87	6,04	140
Max	51,1	18,39	8,91	178
CV	12,89%	32,76%	31,08%	19,74%
Pr > (F)	0,0196*	0,373ns	0,118ns	0,0715ns

PMG= Thousand-seed weight; **Pmoy**.Cap=Headstock **weight**; PGr/Cap=Headstock seed weight; NbGr/Cap=Number of seeds per head; CV=Coefficient of variation; ns=not significant at the 5% level; *=significant at the 5% level; Pr=probability at the 5% level; g=gram; Min=minimum; **Max=maximum.**

III.1.3.Relationship between variables

The correlation test was carried out on the fourteen (14) quantitative variables, which include: average plant height at 30^{e} and 60^{e} days after sowing (Ha30 and Ha60), average plant density at 15^{e} , 30^{e} and 60^{e} days after sowing (Dens15, Dens30, Dens60), duration of the sowing-50% flowering cycle (DSF), duration of the sowing-50% maturity cycle (DSM), seed yield (RdtGr), biomass yield (RdtBiom) and flower head yield (RdtCap), thousand-seed weight (PMG), average flower head weight (PmoyCap), seed weight per flower head (PGr/Cap) and number of seeds per flower head (NbGr/Cap). Principal component analysis and hierarchical ascending classification were used to see how the variables behaved as a function of different fertiliser doses.

➢ Correlation between quantitative variables

Flower head yield was strongly correlated (r=0.95) with seed yield. Seed yield was significantly and positively correlated with biomass yield, thousand-seed weight, average seedhead weight, seed weight per seedhead, number of seeds per seedhead and average plant height. Plant density, on the other hand, was negatively correlated with the yield parameter and its components, and with mean plant height. However, density at 30^{e} and 60^{e} JAS were the two most correlated variables (r=0.99). The phenological parameters were weakly or negatively correlated with all the other variables (**Figure 13**).

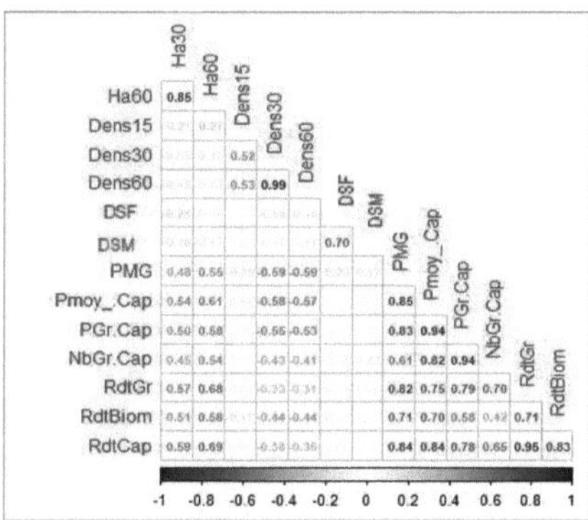

Figure 13: Pearson correlation matrix

➢ Principal component analysis (PCA)

Principal component analysis (PCA) was carried out on the fourteen (14) initial parameters. After examining the eigenvalue and the proportion of variance, the first two components (Dim1 and Dim2) were retained, explaining 78.1% of the variability observed (**Table 12)**. The first component (Dim1) explains 51.5% of the variability. It is defined by variables such as thousand-seed weight (TSP), average weight of a flower head (Pmoy.Cap), plant height at 60^{e} JAS (Ha60), seed yield (RdtGr), seed weight per flower head (PGr/Cap) and flower head yield (RdtCap), each of which contributes more than 10%. The second component (Dim2) explains 26.6% of the variability. The variables contributing to the definition of this dimension are: plant density at 15^{e} JAS (Dens15), duration of the sowing-50% flowering cycle (DSF), plant density at 60^{e} JAS (Dens60), duration of the sowing-50% maturity cycle

(DSM), plant height at 30^{e} JAS (Ha30) and biomass yield (RdtBiom). Phenological variables and biomass yield, although visible on the second component, are best represented by the third dimension (Dim3). The contribution of all the variables to the first two components (Dim1 and Dim2) is between 3 and 9% **(Figure 14)**.

Table 12: Contribution of variables to principal components

Main components	Dim1	Dim2	Dim3	Dim2+Dim1
Eigenvalue	7, 21	3, 73	1, 59	10,94
Variance	51, 5	26, 6	11,3	78,1
Cumulative variance	51, 5	78,1	89,4	78,1
Contributions (%)				
Ha30	7, 63	8,66	0,50	7,98
Ha60	10,69	3,69	0,00	8,31
Dens15	0,11	17,24	1,27	5,95
Dens30	7,74	8,36	6,52	7,95
Dens60	6,63	9,73	7,25	7,68
DFS	0,27	10,79	31,25	3,86
DSM	0,00	9,71	33,49	3,31
PMG	11,74	0,58	0,34	7,94
Pmoy_.Cap	11,28	2,25	0,39	8,20
PGr /Cap	10,46	4,68	0,01	8,49
NbGr/Cap	6,58	7,14	0,00	6,77
RdtGr	10,54	4,63	1,66	8,53
RdtBiom	6,06	7,19	14,58	6,44
RdtCap	10,25	5,31	2,63	8,57

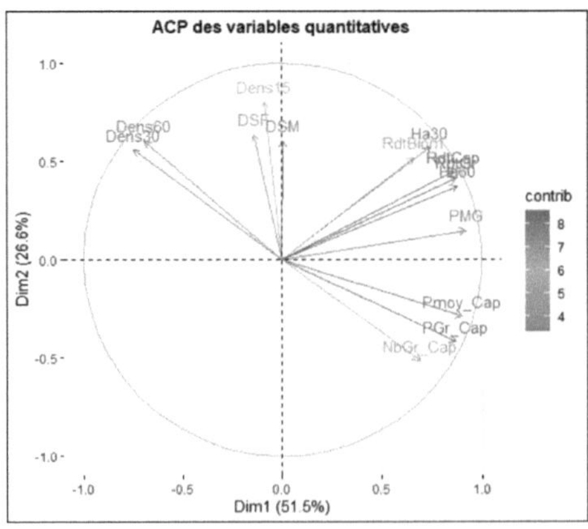

Figure 14: Distribution of variables on the axes

The variables and treatments that are closest to each other on the biplot are positively and strongly correlated. Thus, the biplot clearly shows that :

• treatment F6 is very close to plant height (Ha30 and Ha60), yield (RdtGr, RdtBiom and RdtCap) and thousand-seed weight (PMG), so it is associated with these variables;

• treatments F9, F4 and F8 are associated with yield components (Pmoy.Cap, PGr/Cap and NbGr/Cap);

• density at 30^e and 60^e JAS is the only variable closest to the F2 treatment;

• treatments F1, F3, F7 and F5 are the closest to the plant density at 15^e JAS and phenological variables (**Figure 15**).

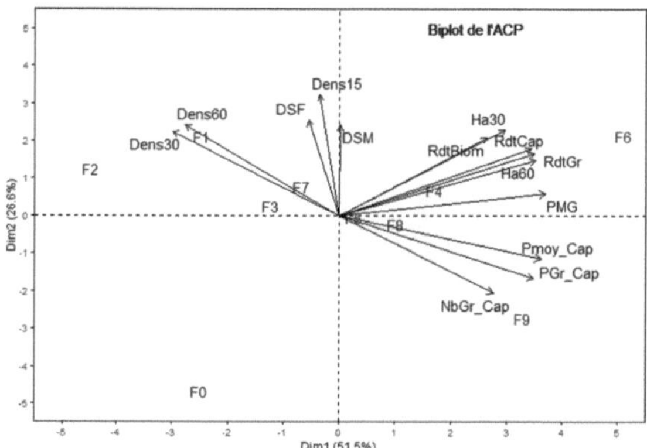

Figure 15: Distribution of the fertilisation plan around the main axes defined by the variables studied

➢ Hierarchical ascending classification

Hierarchical ascending classification was used to discriminate between the different groups of treatments. Five (5) clusters were identified (**Figure 16**):

• **Group 1**: **F2** treatment, in this group all the mean values of the variables are below the general mean except for plant density at 30^e and 60^e JAS;

• **group 2**: this group contains only the control treatment F0 where all variables have mean values below their overall mean;

• **group 3**: with the largest number of treatments (F1, F3, F4, F5, F7 and F8), this group is mixed. In fact, it is characterised by an average plant height at 15^e JAS, an average plant density at 60^e JAS and a seed weight per capitulum lower than the general average. On the other hand, the mean values for plant density at 15^e and 30^e JAS, as well as thousand-seed weight and yield, are all higher than the general average. For the rest of the variables, the values are more or less equal to the overall average for all treatments;

• **group 4**: comprising treatment F9, this group is characterised by average plant density values that are lower than the overall average, while those for height and yield parameters and its components are higher than the average for all treatments; and

• **group 5**: consisting solely of treatment F6, this is the group in which all the mean values of the variables are higher than the overall mean for all the treatments.

In summary, the hierarchical ascending classification showed that mineral fertilisation (F1 and F2) had a greater influence on growth variables than on yield. On the other hand, organo-mineral fertilisation (F6, F7, F8 and F9) had a positive impact on both, but with a more significant effect on yield. F6 dominated all the other treatments. It performed best in terms of both growth and yield. Organic fertilisation (F3 and F4) appears to have a mixed effect. Sometimes it contributes to growth, sometimes to yield.

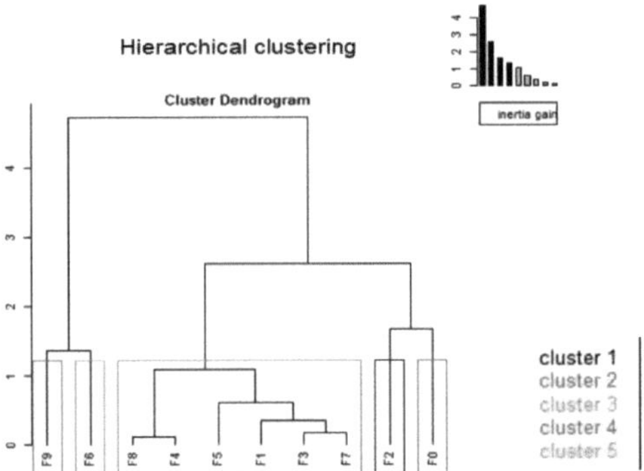

Figure 16: Hierarchical ascending classification of different treatment groups

III.2. DISCUSSION

III.2.1. Effect of fertilisation on growth and development parameters

➢ **Plant height**

The ANOVA result disagrees with those obtained by Ahmad et al. (2018) and Yerima et al. (2014), who demonstrated that Hybrid and African Giant sunflower varieties showed significant differences in plant height as a function of fertilisation rates applied. However, the mean values obtained are very similar to those reported by Yerima et al. (2014). This similarity could be explained by the fact that plant height generally depends on genetics, environmental conditions and cultivation practices. According to Jarecki (2022), plants that have received higher nitrogen fertilisation achieve greater height, while those fertilised at lower doses show significantly reduced growth.

➢ Plant density

The ANOVA results agree with those of Jarecki (2022), who showed that pre-harvest plant density is not affected by fertilisation. Density levels varied between plots and dates, which could be attributed to the emergence and germination problems observed after the first sowing. In fact, the emergence rate rose from 26% at first sowing to 64% at resowing. These difficulties, which have a negative impact on density, are probably linked either to the heavy rainfall recorded before and after sowing, or to inadequate seedbed preparation. These observations are in line with those of SODEFITEX (2005), which found that heavy rainfall after sowing in the agro-ecological region of eastern Senegal could adversely affect emergence. Similarly, Joseph et al (2007) reported that poor rainfall conditions made sunflower germination and emergence difficult in the cotton-growing zone of northern Cameroon. Growing sunflowers at high density can limit the moisture and nutrients available during seed filling, while low density can lead to under-utilisation of resources and an increased risk of weeds. It is important to adjust plant density according to specific soil and climatic conditions. The A.I. Baraev Scientific and Grain Production Centre recommends having 25,000 to 40,000 plants per hectare in periods of low humidity, and increasing the sowing rate by 25% to compensate for germination problems, pests and soil diseases (Kaskarbaev et al., 2023).

➢ The length of the cycle sowing-50% flowering and sowing-50% maturity

The analysis of variance did not reveal a significant difference at the 5% threshold, which contrasts with the findings of Ahmad et al. (2017), who observed a significant impact of fertilisation on the maturity date. The average values for the sowing-50% flowering and sowing-50% maturity cycles correspond to those reported by SODEFITEX (2005), which found, during two experimental campaigns in eastern Senegal, that the 50% flowering date for sunflower is between 40 and 55 days. However, the earliest flowering and ripening dates were recorded in plots that had received organic and organo-mineral fertilisation (F5, F6, F7, F8 and F9). This slight advance in flowering and maturity, compared with plots that received only mineral fertilisation, could be explained by the combined input of nutrients (N, P, K, trace elements) and organic matter, via fertilisers (NPK triple 15 and urea) and compost. The decomposition and mineralisation of organic matter releases essential nutrients in soil that is poor in organic matter, nitrogen, phosphorus, potassium and exchangeable bases (**Table 1**). Organic matter, provided by compost ('Toss Gui'), and fertilisers improve the physical, chemical and biological properties of the soil. These results confirm the findings of Yerima et al (2014), who showed that the Italian White sunflower variety flowers more quickly when the amount of hen droppings (organic matter of animal origin) is increased up to 4.2 tonnes per hectare.

III.2.2. Effect of fertilisation on yield and its components

➢ **Performance**

The ANOVA results are consistent with those reported by Ali et al. (2012), Nasim et al. (2012) and Sincik et al. (2013) who also demonstrated that seed yield of sunflower (hybrid variety) is significantly affected by nutrient supply. On the other hand, Ali et al. (2012) showed that increasing rates of nitrogen could have a positive effect on the number of flower heads and the number of seeds per flower head.

Although the study was conducted on an experimental station, the yields obtained remain below the world average (18.5 $q.ha^{-1}$ according to FAOSTAT, 2023). Several factors could explain this drop, including limitations in crop establishment, soil preparation, late sowing, inadequate plant health monitoring and late harvesting. In a study by Kulygin et al (2017), it was observed that sunflower response to fertiliser application varied according to tillage type, leading to a yield increase of 10.6-18.7% compared to control plots. Sunflower yield under rainfed conditions is mainly affected by irregular and insufficient rainfall, despite its better adaptation to water stress compared with other crops (Gordeyeva et al., 2024; Ahmad et al., 2022; Fatahi et al., 2022 and Hlisnikovský et al., 2016). In addition to the major nutrients (N, P, K) provided by mineral fertilisation, organic and organo-mineral fertilisation has enriched the soil with trace elements such as boron, molybdenum, zinc and iron, as well as organic matter via 'Toss Gui' compost. The C/N ratio of this compost (16 to 22) encourages rapid mineralisation of nitrogen, making this nutrient available to the plant throughout its growth. This increased availability could explain the yield differences observed between plots that received organo-mineral or organic fertilisation and those that received only mineral fertilisation or no fertilisation at all (controls). This observation is supported by several studies. Sefaoglu et al. (2021) obtained the highest seed yields by combining nitrogen and vermicompost. Ahmad et al (2018) confirmed that nitrogen fertilisation had a positive effect on sunflower growth and development. Hajduk et al. (2017) observed that increasing mineral fertilisation with nitrogen reduced the boron content of sunflower biomass. As a result, Alves et al. (2017) concluded that higher rates of nitrogen should be combined with boron fertilisation to achieve significantly higher sunflower yields. Studies have shown that nitrogen sources and doses significantly influence sunflower yield and agronomic traits, particularly the use of ammonium nitrate, which improves plant height, flower head diameter, thousand-seed weight and overall yield (Boldisov and Bushnev, 2016 and Osama et al., 2010).

➢ **Performance components**

The ANOVA results are in agreement with Ahmad et al. (2018), who demonstrated that increasing NPK levels significantly increased grain yield and thousand-seed weight. In contrast, Hlisnikovský et al. (2016) found that thousand-seed weight, total seed weight, number of seeds per flower head and yield per hectare were not significantly affected by fertiliser treatment.

The plots that received organic and organo-mineral fertilisation performed best in terms of yield components, outperforming those fertilised only with mineral fertilisers or not fertilised at all. This difference could be attributed to the trace elements present in the compost. These results are corroborated by Steiner and Zoz (2015), Pattanayak et al. (2017), and Andrade et al. (2021), who reported that the combination of NPK and trace elements (such as zinc, boron and molybdenum) promotes plant growth and improves sunflower yield. Al-Amery et al (2011) also showed that boron fertilisation reduced the number of empty seeds, thereby significantly increasing yield and its components. Kandil et al. (2017) showed that a high dose of nitrogen increased plant height, stem thickness, the number of leaves per plant, leaf area, the number of seeds, and so on. per flower head, flower head diameter and thousand-seed weight. However, mean temperatures above 30°C and potentially inadequate supplementary irrigation during flowering could have an impact on yield and its components. Rondanini et al (2006) report that a mean temperature above 30°C for four days or more can lead to yield loss by reducing grain weight and increasing the proportion of very small grains. According to Alberio et al (2015) and Ştefan et al (2022), drought-related water stress affects plant morphological characteristics, such as reduced leaf area, and also impacts yield characteristics, including the number of seeds per flower head and thousand-seed weight. The period of highest drought sensitivity for number of seeds per square metre and average seed weight is around 40 days around flowering. Nouri et al (2011) indicate that a 50% reduction in humidity during flowering results in a decrease of more than 30% in the number of seeds and 20% in the average seed weight.

III.2.3. Relationship between variables

- **Correlation between quantitative variables**

The results of the correlation analysis showed that the majority of yield-related traits have a significant influence on sunflower seed yield. With the exception of the length of the sowing-50% maturity cycle, the length of the sowing-50% flowering cycle and plant density, all the other variables studied were positively correlated with seed yield. These results are in line with those of Ahmad et al (1991), Kaya et al (2007) and Machikowa & Saetang (2008). Due to nutrient competition, increasing the density of sunflower plants resulted in a decrease in grain weight, thousand-seed weight, number of grains per capitulum and weight per capitulum. These observations are consistent with those of Li et al. (2019).

- **Principal component analysis and hierarchical ascending classification**

The results of the principal component analysis and the hierarchical ascending classification suggest that the efficiency of mineral elements in plant metabolism and their role in meristematic activity are responsible for the increase in plant height. These elements promote cell division, elongation and the formation of vegetative organs, leading to more vigorous growth. This improved growth leads to an increased accumulation of photosynthesis products, resulting in a higher seed weight. In addition, several studies on the response of sunflower to nitrogen and phosphorus fertiliser levels, conducted by Al-Thabet (2006), Eba & MMM (2010), Ali et al. (2012), Ali et al. (2014), and Li et al. (2018), have shown that the application of nitrogen and phosphorus significantly improves growth and yield.

CONCLUSION

This study assessed the effects of organo-mineral fertilisation on sunflower growth and productivity under rainfed conditions in the southern groundnut basin. More specifically, it assessed the effect of different fertilisation plans on sunflower growth and productivity in the region.The organo-mineral fertilisation treatments F6 and F9 showed the best performance in terms of yield. Treatment F6 recorded a seed yield of 738.5 ± 199.7 $kg.ha^{-1}$, while treatment F9 obtained 569.5 ± 242.0 $kg.ha^{-1}$. For biomass yield, F6 produced 1140.2 ± 262.8 $kg.ha^{-1}$, and F9 achieved 721.9 ± 334.3 $kg.ha^{-1}$. Head yield was 1504.8 ± 390.3 $kg.ha^{-1}$ for F6 and 1172.9 ± 461.3 $kg.ha^{-1}$ for F9. The highest thousand-seed weight, at 51.1 ± 4.6 g, was observed with the F6 treatment, while the F9 treatment showed the highest values for the average weight of a flower head (18.39 ± 6.26 g) and the weight of seeds per flower head (8.91 ± 3.25 g). The tallest plants were measured in treatments F6 (90.5 ± 5.5 cm) and F9 (87.0 ± 9.1 cm). Principal component analysis (PCA) revealed significant differentiation between treatments. Treatment F6 was associated with plant height, yield variables and thousand-seed weight. Treatment F9 was linked to the average weight of a flower head and the weight of seeds per flower head. Hierarchical ascending classification (HAC) grouped the treatments into distinct classes, revealing notable similarities and differences between the groups in terms of sunflower growth and yield, with treatment F6 having the highest values, followed by treatment F9. In conclusion, this study makes a significant contribution to our understanding of the effect of organo-mineral fertilisation on sunflower cultivation, and offers avenues for the sustainable improvement of agricultural practices in the groundnut basin.

Despite these promising results, several aspects require further research to consolidate these conclusions. We therefore make the following recommendations:

- Repeat the trial under different conditions to confirm or refute the results obtained by studying variety*environment*management interactions;
- test other fertilisation formulas in combination with organic matter ;
- to study the effect of the fertilisation plan on the chemical and biological properties of the soil in order to promote sustainable soil fertility management.

REFERENCES

Abga, P. T. (2013). Determination of organo-mineral fertilisation and sowing density options for intensification of maize production in the eastern region of Burkina Faso. Rural development engineer dissertation.

Agreste (2014). Enquête Pratiques culturales 2011. Agreste. Les Dossiers- n°21-July 2014. 70p.

Agridea (2007). Agridea - Sunflower.

AGROPOL. (2023). Press release on the cooperation project for the development of oilseed and protein crop sectors in Senegal. Paris. 2 March 2023. 1p.

Ahmed, N. A. K. (2022). Evaluation of ecosystem services and dis-services provided by multiservice intermediate crops and biofumigation to improve sunflower productivity (Doctoral dissertation, Institut National Polytechnique de Toulouse-INPT).

Ahmad, H. M., Wang, X., Fiaz, S., Azeem, F., & Shaheen, T. (2021). Morphological and physiological response of Helianthus annuus L. to drought stress and correlation of wax contents for drought tolerance traits. Arabian Journal for Science and Engineering, 1-15.

Ahmad, M.I., Ali, A., He, L., Latif, A., Abbas, A., Ahmad, J., Ahmad, M.Z., Asghar, W., Bilal, M. & Mahmood, M.T. (2018). Nitrogen effects on sunflower growth: A review. Int. J. Biosci, 12, 91-101.

Ahmad, Q., Rana, M.A. & Siddiqui, S.U.H. (1991). Sunflower seed yield as influenced by some agronomic and seed characters. Euphytica, 56(2), 137-142.

Akanza, K. P., Sanogo, S., & N'Da, H. A. (2016). Combined influence of organic and mineral manures on maize nutrition and yield: impact on soil deficiency diagnosis. Tropicultura, 34(2), 208- 220.

Akanza, P. K., & Yoro, G. (2003). Synergistic effects of mineral fertilizers and poultry manure in improving the fertility of a ferrallitic soil in western Côte d'Ivoire. Agronomie Africaine, 15(3), 135-144.

Alberio, C., Izquierdo, NG, & Aguirrezábal, LAN (2015). Physiology and agronomy of sunflower crops. In Sunflower (pp. 53-91). AOCS Press.

Al-Amery, M. M., Hamza, J. H., & Fuller, M. P. (2011). Effect of boron foliar application on reproductive growth of sunflower (Helianthus annuus L.). International Journal of Agronomy, 2011(1), 230712.

Al-Thabet, S. S. (2006). Effect of plant spacing and nitrogen levels on growth and yield of sunflower (Helianthus annus L.). J. King Saud Univ. Agric. Sci, 19(1), 1-11.

Ali, A., Ahmad, A., Khaliq, T., Afzal, M., Iqbal, Z., & Qamar, R. (2014). Plant population and nitrogen effects on achene yield and quality of sunflower (Helianthus annuus L.) hybrids. In International Conference on Agricultural, Environmental and Biological Sciences (AEBS-2014) April (pp. 24-25).

Ali, A., Ahmad, A., Khaliq, T., Afzal, M., & Iqbal, Z. (2012). Achene yield and quality response of sunflower hybrids to nitrogen at varying planting densities. In International Conference on Agriculture, Chemical and Environmental Sciences (ICACES'2012) Oct (pp. 6-7).

Alves, L. S., Stark, E. M. L. M., Zonta, E., Fernandes, M. S., Santos, A. M. D., & Souza, S.
R. D. (2017). Different nitrogen and boron levels influence the grain production and oil content of a sunflower cultivar. Acta Scientiarum. Agronomy, 39(1), 59-66.

Andrade, A.F., Vicosi, K.A., Bueno, A.M., Flores, R.A., Santos, C.L.R., Santos, G.G. & Mesquita, M. (2021). Nutritional and biomass aspects of Helianthus annuus according to boron application in the soil. Australian Journal of Crop Science, 15, 899-908.

ANSD. (2023). Bulletin mensuel des statistiques du commerce extérieur. Dakar: ANSD ,12, 59p.

Beye, G., Ndione, J-A., Faye, A., Sall, A.B. & Ndiaye, D.S. (2012). Analyse des potentialités agricoles et pastorales du département de Nioro du Rip, Dakar: CSE, 37 Slides.

Bonjean, A. (1993). Le tournesol : économie, origine, histoire, écologie, sélection. Les Editions de l'Environnement. 242 p.

Bonjean, A. & Pham-Delegue, M.H. (1986). Sunflower pollination. OPIDA 10.8.11, (99), 212-218.

Bret-Mestries, E., Debaeke, P., Seassau, C., Dechamp-Guillaume, G., Langlade, N., Aubertot, J. N., & Casadebaig, P. (2016). Dossier Tournesol, 10 years of collaborative research-Marked results.

Bureau Pédologie du Sénégal (B.P.S). (1993). Semi-detailed study of the soils of Nioro. 123p.

Burke, J. M., Tang, S., Knapp, S. J., & Rieseberg, L. H. (2002). Genetic analysis of sunflower domestication. Genetics, 161(3), 1257-1267.

CETIOM. (2004). Benchmark stages of sunflower. In: CETIOM (Ed.), Guide de l'expérimentateur tournesol. Available at: http://www.cetiom.fr/tournesol/cultiver-du-tournesol/atouts- points-cles/stadesreperes/ (consulted on 10/01/2024).

CETIOM. (1999). Sunflower in 99: cultivation techniques and economic context.

Chandrasekaran, B., Annadurai, K. & Somasundaram, E. (2010). Sunflower in Agronomy of field crops and biofuel plants, New Age International (P) Ltd, Publishers. p.594-597.

CNRA. (2023). Données de sol essais tournesol Nioro HIV 2023. Soil-water-plant laboratory

Dembele, I. (1994). Production and use of organic manure. Fiche synthétique d'information. 19p.

Demol, J., Baudoin, J. P., Louant, P. B., Marechal, R., Mergeai, G., & Otoul, E. (2002). Plant improvement. Application to the main species grown in tropical regions. Les presses agronomiques de Gembloux. : 213-221.

Dieng, M. (2021). Effects of different doses of organo-mineral fertilization on soil chemical properties, growth and yield of rice (Oryza sativa L.) at Balmadou (Casamance- Senegal). 61p.

EBA, O., & MMM, A. (2010). Response of sunflower (Helianthus annuus L.) to phosphorus and nitrogen fertilization under different plant spacing at new valley.

Ebrahimi, A. (2008). Contrôle génétique de la qualité des graines chez le tournesol (Helianthus annuus L.) soumis à la sécheresse (Doctoral dissertation, Institut National Polytechnique (Toulouse)).177p.

El Asri, M., Kayaf, M., & El Fachtali, M. (2002(a)). Effect of early water stress during the reproductive phase of sunflower. Proceedings of the First Symposium on the Development of the Oilseed Sector in Morocco. Société Marocaine d'Agronomie (SMA), 211-215.

El Asri, M., Bamouh, A., Tahiri, I. & Kayaf, M. (2002(b)). Controlling the distribution of irrigation water between the vegetative and reproductive phases. Institut Agronomique et

Vétérinaire Hassan II.

El-Hassan, O. M., Ali, E. A., Mohamed, A. E., & Mohamed, M. Y. (2007). Response of sunflower to different types of fertilizer and nitrogen levels in the rain-fed area of the Blue Nile State.

Escoffier, I. & Gipouloux, M. (2022). Fertilising sunflowers in organic farming. La France Agricole, 3948, Available at www.lafranceagricole.fr (consulted on 26/02/2024).

Evon, P. (2008). New biorefinery process for whole plant sunflower by thermo-mechanical-chemical fractionation in a twin-screw extruder: study of the aqueous extraction of lipids and the shaping of the raffinate into agromaterials by thermomoulding (Doctoral dissertation).

Fairhurst, T. (2015). Integrated soil fertility management manual.176p.

Falisse, A., & Lambert, J. (1994). Mineral and organic fertilisation. Agronomie Moderne, Bases Physiologiques et Agronomiques de la Production Végétale (EL HASSANI, TA et al., Eds.), Hatier, AUPELF-UREF, Paris, 377-398.

FAOSTAT (2023). Production. Crops and animal products. Primary crops & derived crops. Sunflower. Available at http://www.fao.org/faostat/fr/#data (consulted on 16/02/2024).

Fatahi, A., Safarian Zengir, V., Sobhani, B., Kianian, M., & Ghahremani, A. (2022). Assessment and zoning of suitable climate for economic development of cultivation of Sunflower (Helianthus annuus) garden crop (Ardabil Province, Iran). European Journal of Horticultural Science, 87(1), 1-12.

Faye, B. A. (2022). Effects of cropping method and organo-mineral fertilisation on soil chemical properties, growth and yield of rice (Oryza sativa L.) in acid sulphate soils in Basse-Casamance.

Fertial. (2017). Fertilizer use manual. 128p.

Fried, G., Le Corre, V., Rakotoson, T., Buchmann, J., Felten, E. & Chauvel, B. (2022). Consequences of the use of herbicide-tolerant sunflower varieties on the flora of agrosystems. Agronomie, Environnement & Sociétés, 12(1), art-17

Fougeroux, A., Leylavergne, S., Guillemard, V., Geist, O., Gary, P., Cenier, C., Caumes-Sudre, E., Senechal, C. & Vaissière, B. (2017). Effect of insect pollinator activity on pollination and yield of sunflower for consumption. OCL, 24(6): D603.

Gaye, A. T., Lo, H. M., Sakho-Djimbira, S., Fall, M. S. & Ndiaye, I. (2015). Senegal:

Review of the socio-economic, political and environmental context. Centre de suivi écologique: Dakar.

Gordeyeva, Y., Shelia, V., Shestakova, N., Amantayev, B., Kipshakbayeva, G., Shvidchenko, V., ... & Hoogenboom, G. (2023). Sunflower (Helianthus annuus) Yield and Yield Components for Various Agricultural Practices (Sowing Date, Seeding Rate, Fertilization) for Steppe and Dry Steppe Growing Conditions. Agronomy, 14(1), 36.

Hajduk, E., Gąsior, J., Właśniewski, S., Nazarkiewicz, M., & Kaniuczak, J. (2017). Influence of liming and mineral fertilization on the yield and boron content of potato tubers (Solanum tuberosum L.) and green mass of fodder sunflower (Helianthus annuus L.) cultivated in loess soil. Journal of Elementology, 22(2).

Heiser, C. B. (1985). Some botanical considerations of the early domesticated plants north of Mexico. Prehistoric food production in North America, 57-72.

Heiser, C. B., Smith, D. M., Clevenger, S. B. & Martin, W. C. (1969). The north american sunflowers (Helianthus). Memoirs of the Torrey Botanical Club, 22(3), 1-218.

Hlisnikovský, L., Kunzová, E., Hejcman, M., Škarpa, P., & Menšík, L. (2016). Effect of nitrogen, boron, zinc and molybdenum application on yield of sunflower (Helianthus annuus L.) on greyic phaeozem in the Czech Republic. Helia, 39(64), 91-111. https://doi.org/10.1515/helia-2015- 0011

Ibrahim, H. M. (2012). Response of some sunflower hybrids to different levels of plant density. Apcbee Procedia, 4, 175-182.

INA P-G. (2003). Fiche sur le tournesol. AGER Department. 21p.

ISRA. (1998). Rapport d'activités. Rapport annuel-CNRA-1998. Senegal, 124 p.

CNRA Bambey (2023). Nioro Protocole Essai Fertilisation Tournesol Hiv 2023. 5p.

Jarecki, W. (2022). Effect of varying nitrogen and micronutrient fertilization on yield quantity and quality of sunflower (Helianthus annuus L.) achenes. Agronomy, 12(10), 2352. https://doi.org/10.3390/agronomy12102352

Joseph, W., Saïdou, I. (2007). Feasibility study of soybean and sunflower in the cotton zone of North Cameroon. Results of the 2006 experimental season. Institut de Recherche Agricole pour le Développement, Cameroon. 55p.

Kandil, A. A., Sharief, A. E., & Odam, A. M. A. (2017). Response of some sunflower hybrids (Helianthus annuus L.) to different nitrogen fertilizer rates and plant densities. International Journal of Environment, Agriculture and Biotechnology, 2(6), 238990.

Kaskarbaev, J., Pokhorukov, Y., Kitdralina, A., Sasykov, A., Werner, A. (2023). Technology of cultivation of oilseeds in the North of Kazakhstan. A.I. Baraev Research and Production Center for Grain Farming. Available online: https://baraev.kz/statya/427-tehnologiya-vozdelyvaniya- maslichnyh-kultur-na-severe-kazahstana.html?ysclid=lepbd5nao0680069173y.

Kaya, Y., Evci, G., Durak, S., Pekcan, V., & Gücer, T. (2007). Determining the relationships between yield and yield attributes in sunflower. Turkish Journal of Agriculture and Forestry, 31(4), 237-244.

Killi, F. A. T. I. H. (2004). Influence of different nitrogen levels on productivity of oilseed and confection sunflowers (Helianthus annuus L.) under varying plant populations. International Journal of Agriculture and Biology, 4, 594-598.

Leclercq, P., Cauderon, Y., & Dauge, M. (1970). Selection for mildew resistance in sunflower from Topinambour x Sunflower hybrids. Ann Amél Plantes, 20(3), 363-373.

Lecomte, V. & Nolot, J.M. (2011). Place du tournesol dans le système de culture, Innovations Agronomique, 14, 59-76.

Lecomte, J. (1962). Observations on the pollination of sunflowers (Helianthus annuus L.). Les Annales de l'Abeille, 5(1), 69-73.

Lepennetier, A. (2015). Partial resistance to sunflower phoma: epidemiological approach and measurement of resistance components (Doctoral dissertation, University of Lorraine).

Li, J.; Qu, Z., Chen, J., Yang, B., & Huang, Y. (2019). Effect of Planting Density on the Growth and Yield of Sunflower under Mulched Drip Irrigation. Water, 11(4), 752. https://doi.org/10.3390/w11040752.

Li, S. T., Yu, D. U. A. N., Guo, T. W., Zhang, P. L., Ping, H. E., & Majumdar, K. (2018). Sunflower response to potassium fertilization and nutrient requirement estimation. Journal of integrative agriculture, 17(12), 2802-2812. https://doi.org/10.1016/S2095-3119(18)62074-X.

Linz, G. M., & Hanzel, J. J. (1997). Birds and sunflower. Sunflower technology and production, 35, 381-394.

Maertens, C., & Bosc, M. (1981). Study of the evolution of sunflower rooting (Stadium variety).

Machikowa, T & Saetang, C (2008). Correlation and path coefficient analysis on seed yield in sunflower. Suranaree Journal of Science Technology, 15(3), 243-248.

Mas seeds (2024). Sunflower essentials. A technical guide to successful growing.44p. Available on: click here

Mazoyer, M. (2002). Larousse agricole: Le monde paysan au XXIe siècle. Paris: Larousse, 695- 697.

Merrien, A. (1986). Physiology of sunflower. Cahier Technique Tournesol, Ed. CETIOM, 47 p.

Mojiri, A., & Arzani, A. (2003). Effects of nitrogen rate and plant density on yield and yield components of sunflower.

Naima, M., & Imane, M. (2013). Contribution to the bibliographic study on fertilizers and their analyses (Doctoral dissertation, Faculty of Science and Technology).

Ndiaye, A., Ndiaye, O., Bamba, B. & Gueye, M. (2019). Effects of organo-mineral fertilisation on growth and yield of sanio millet (Pennisetum glaucum LR Br) in Upper Casamance (Senegal). European Scientific Journal November, 15(33), 1857- 7431.

Ndiaye, A., & Hereau, G. (2012). Agronomy course-3ème year (ENSA).

Nouri, L., Ykhlef, N., & Sarrafi, A. (2011). Identification of physiological markers of drought tolerance in sunflower.

Nsome, P.A. (1999). Régénération des sols dégradés dans le Bassin arachidier : optimisation de l'eau et des éléments nutritifs du maïs, dissertation. ENCR.

OECD (2006), "Section 9 - Sunflower (Helianthus Annuus L.)", in Safety Assessment of Transgenic Organisms, Volume 1: OECD Consensus Documents, OECD Publishing, Paris. Available at: https://doi.org/10.1787/9789264095380-12-e (consulted on 05/01/2024).

Osama, M., Elhassan, A., Elnaiem, A., Mohamed, A. A. E., & Mohamed, M. Y. (2010). Response of Sunflower to different types of fertilizer and nitrogen levels in the rain-fed area of the Blue Nile State. Agricultural Research Corporatio Unit, Wad Medani, 26-34.

Ouandaogo, N., Ouattara, B., Pouya, M. B., Gnankambary, Z., Nacro, H. B., & Sedogo, P.

M. (2016). Effects of organomineral manures and crop rotations on soil quality. International Journal of Biological and Chemical Sciences, 10(2), 904-918.

Ouédraogo, E., Mando, A., & Zombré, N. P. (2001). Use of compost to improve soil properties and crop productivity under low input agricultural system in West Africa. Agriculture, ecosystems & environment, 84(3), 259-266.

Parker, F. D. (1981). Sunflower pollination: abundance, diversity, and seasonality of bees on male-sterile and male-fertile cultivars. Environmental Entomology, 10(6), 1012-1017.

Pattanayak, S., Behera, A. K., Das, P., Nayak, M. R., Jena, S. N., & Behera, S. (2017). Performance of summer sunflower (Helianthus annuus L.) hybrids under different nutrient management practices in coastal Odisha. Journal of Applied and Natural Science, 9(1), 435-440.

Petit, J., & Jobin, P. (2005). La fertilisation organique des cultures: les bases. Fédération d'agriculture biologique du Québec.

Pieri C. (1989). Fertilité des terres de savanes. Thirty years of agricultural research and development south of the Sahara. Paris, Ministère de la Coopération et du Développement, CIRAD-Irat, La Documentation française. 444 pp.

Putnam, D. H., Oplinger, E. S., Hicks, D. R., Durgan, B. R., Noetzel, D. M., Meronuck, R. A., & Schulte, E. E. (1990). Sunflower. Alternative Field Crops Manual.

Pressenza (2019). Senegal: organic farming at the service of society. Available at https://www.pressenza.com/fr/2019/07/senegal-lagriculture-biologique-au-service-de-la-societe/ (consulted on 18/01/2024).

Ramde, R. (2014). Physical and biochemical characterization of 12 varieties and 16 cultivars of sunflower (Helianthus annuus. L). End of cycle dissertation. Université Polytechnique de Bobo- Dioulasso, Burkina Faso.

Rondanini, D., Mantese, A., Savin, R., & Hall, A. J. (2006). Responses of sunflower yield and grain quality to alternating day/night high temperature regimes during grain filling: effects of timing, duration and intensity of exposure to stress. Field Crops Research, 96(1), 48- 62.https://doi.org/10.1016/j.fcr.2005.05.006

Rollier, M. (1972). Nutrient requirements of sunflower. In Proceeding 5th International sunflower conference, Clermont Ferrand, France (pp. p73-76).

SEDAB/BIOTOSS. Document flyers: the process of making and using organic fertiliser.

Sefaoğlu, F., Ozturk, H., Oztürk, E., Sezek, M., Toktay, Z. & Polat, T. (2021). Effect of organic and inorganic fertilizers or their combinations on yield and quality components of oil seed sunflower in a semi-arid environment. Turkish Journal of Field Crops, 26(1), 88-95.

Seiler, G. J. (1997). Anatomy and morphology of sunflower. Sunflower technology and production, 35, 67-111.

Siboukeur, A. (2013). Appreciationof the value fertilizing of different types of manure (Doctoral dissertation, Université Kasdi Merbah-Ouargla).

Sincik, M., Goksoy, A. T., & Dogan, R. (2013). Responses of sunflower (Helianthus annuus L.) to irrigation and nitrogen fertilization rates.

Soltner, D. (2011). The basics of crop production. Tome I. The soil and its improvement. Collection Sciences et Techniques Agricoles, 23rd. Ed. Paris, 472p.

Soltner, D. (2005). Les grandes productions végétales. Sciences et techniques agricoles.

Somda, B. B., Ouattara, B., Serme, I., Pouya, M. B., Lompo, F., Taonda, S. J. B., & Sedogo, P. M. (2017). Determination of optimal doses of organo-mineral manures in microdose in the Sudano-Sahelian zone of Burkina Faso. International Journal of Biological and Chemical Sciences, 11(2), 670-683.

SODEFITEX. (2011). Assessment of sunflower cultivation in Senegal

SODEFITEX. (2005). Sunflower experimentation in Senegal (2005 cropping season), 11p.

Steiner, F., & Zoz, T. (2015). Foliar application of molybdenum improves nitrogen uptake and yield of sunflower. African Journal of Agricultural Research, 10(17), 1923-1928.

Ştefan, I. O., & Constantinescu, E. (2022). Research on the behavior of some sunflower cultivations in the specific conditions of the North Area of Olt County. Annals of the University of Craiova-Agriculture Montanology Cadastre Series, 52(2), 170-173.

Süzer, S. (2010). Effects of nitrogen and plant density on dwarf sunflower hybrids. Helia, 33(53), 207-214.

Syngenta (2013). The sunflower yield crop guide -NK Seeds.

Temagoult, M. (2009). Analysis of the variability of the response to water stress in recombinant lines of Sunflower (Helianthus annus L.). Diplôme de magistère en Biotechnologies Végétales. Mentouri University, Constantine.

Terres Univia and Terres Inovia. (2022). Seed quality. Harvest 2022

Terre Inovia and Lidea (2023) Veronika. Fiche variétal du tournesol.

Terres Inovia. (2023(a)). Guide de culture du tournesol. Available at www.terresinovia.fr (accessed 05/01/2023).

Terres Inovia. (2023(b)). Orobanche cumana: using solutions adapted to your situation. Available at: https://www.terresinovia.fr/-/orobanche-cumana-utiliser-des-solutions-adaptees-a- votre-situation (consulted on 23/06/2024).

Terres Inovia. (2020). The true-false facts about sunflower irrigation [WWW Document]. Available on : Irrigation du tournesol : faire une croix sur les idées reçues (terre-net.fr) (accessed 22/06/2024).

Thebaud, V. and Scheiner, J. (2010). One of the keys to growing sunflowers: root development. Expert opinion. Purpan Engineering School, Toulouse, France.

Tounkara, A., Sarr, S., Ndiaye, M., Senghor, Y. & Camara, B. (2022). Renewal of soil fertility in millet-based cropping systems in the groundnut basin of Senegal: trends and avenues for improvement. African & Mediterranean Agricultural Journal -Al Awamia (137). p.139-161

Tounkara, A., Clermont-Dauphin, C., Affholder, F., Ndiaye, S., Masse, D., & Cournac, L. (2020). Inorganic fertilizer use efficiency of millet crop increased with organic fertilizer application in rainfed agriculture on smallholdings in central Senegal. Agriculture, Ecosystems & Environment, 294, 106878. https://doi.org/10.1016/j.agee.2020.106878

Vear, F. (1992). Sunflower. Amélioration des espèces végétales cultivées: objectifs et critères de sélection. Paris (France): Inra.

Wajid Nasim, W. N., Ashfaq Ahmad, A. A., Asghari Bano, A. B., Olatinwo, R., Muhammad Usman, M. U., Tasneem Khaliq, T. K., ... & Muzzammil Hussain, M. H. (2012). Effect of nitrogen on yield and oil quality of sunflower (Helianthus annuus L.) hybrids under sub humid conditions of Pakistan.

World Bank National Accounts Data. (2022). Agriculture, forestry, and fishing, value added (% of GDP) - Senegal. Available on Agriculture, forestry and fishing, value added (%

of GDP) - Senegal | Data (worldbank.org) (accessed 21/02/2023).

Xiao, S., Chen, S. Y., Zhao, L. Q., & Wang, G. (2006). Density effects on plant height growth and inequality in sunflower populations. Journal of Integrative Plant Biology, 48(5), 513-519.

Yerima, B. P. K., Tiamgne, A. Y., & Van Ranst, E. (2014). Response of two sunflower (Helianthus sp.) varieties to hen-dung fertilization on a Hapli-Humic Ferralsol at Yongka Western Highlands Research Garden Park (YWHRGP) Nkwen-Bamenda, Cameroon, Central Africa. Tropicultura, 32(4).

Zheljazkov, V. D., Vick, B. A., Baldwin, B. S., Buehring, N., Coker, C., Astatkie, T., & Johnson, B. (2011). Oil productivity and composition of sunflower as a function of hybrid and planting date. Industrial crops and products, 33(2), 537-543.

APPENDIX

Appendix 1: Procedure for collecting data

Data	Variable	Unit	Period	Number
Climate	Air humidity	%	June - October	Per day
	Supplementary irrigation	m3	June - October	Per day (if applicable)
	Rainfall	mm/day	June - October	Per day
	Air temperature	min and max	June - October	Per day
Soil	Grain size: clay, silts, sands (5 fractions)	%	Before tillage	Before setting up the trial: 3 composite samples per station: 0-20 cm depth for the reference situation At trial harvest: 1 sample per variety and per station: 0-20 cm depth Have it analysed at the ISRA soil laboratory (CNRA)
	pH (water)	Value	Before tillage	
	Total nitrogen (N)	%	Before working on the ground	
	Total carbon (C)	%	Before tillage	
	C/N ratio	Calculation	Before working on the ground	
	Assimilable phosphorus	ppm	Before tillage	
	Changeable bases (Ca, Mg, Na, K)	cmol/kg	Before working on the ground	
	Cation exchange capacity (CEC)	cmol/kg	Before tillage	
Phenological	50% lifting time	JAS	From 3 JAS	In each plot useful
	Duration of 50% of flowering	JAS	From 15-20 JAS	In each plot useful
	Duration of 50% maturity	JAS	From 50-60 JAS	In each useful parcel
Agro morphology	Height (length) of plant	cm	30e and 60e JAS	10 plants at random by plot and date
	Density	plant/plot	15e , 30e and 60e JAS	10 plants at random per plot
	Above-ground biomass weight dryer	kg/plot	0 and 21-30 JAR	In each plot useful
Yield and its components	Total number of flower heads per plant	Counting	At harvest time	10 plants at random per plot
	Dry weight of a flower head	Counting	At harvest time	10 plants at random by plot
	Number of seeds per capitulates	Counting	At harvest time	10 plants at random by useful parcel
	Dry weight of seeds	kg/plot	A 10-12% humidity	In each useful parcel
	Weight 1000 seeds	gram	21-30 JAR	5 lots of 1000 seeds per plot

DAS (days after sowing), **DAR** (days after harvest)

Appendix 2: R scripts used to process and analyse data

Stains	Scripts used
Import	Fert_Tournesol<-read.csv("C:/Users/DELL/Documents/Données Tournesol fertilisation.csv", h=T,sep=";",dec=",")
Viewing data	View(Fert_Tournesol)
Nature of the variables in the dataset	str(Fert_Tournesol)
First 6 values of each variable	head(Fert_Tournesol)
Dataset variable names	names(Fert_Tournesol)
Attach variable names	attach(Fert_Tournesol)
Convert to block and dose factors	Block=as.factor(block); Fertilisation dose =as.factor(Dose)
Descriptive statistics	summary(Fert_Tournesol)
Analysis of variance Verification of assumptions Normality Homogeneity of variance	library(agricolae) ; model2<-aov(PMG~Bloc + Dose, data = Fert_Tournesol); summary(model2) residus<-residuals(model2); shapiro.test(residus) bartlett.test(residus~Dose, data= Fert_Tournesol)
Comparison of averages	Turkey2<- HSD.test(model2, "Dose",group=TRUE,console=TRUE)
Fertilisation alone with error bars + graphic representation	library(dplyr); agro.summary2 <- Fert_Tournesol %>%; group_by(Dose) %>% summarise(sd = sd(PMG), PMG = mean(PMG))
Bar plots + error bars	library(ggplot2); ggplot(agro.summary2, aes(x=Dose, y=PMG)) + geom_col(aes(fill = Dose), position = position_dodge(0.8), width = 0.7)+ geom_errorbar(aes(ymin = PMG, ymax = PMG +sd), width = 0.2, position = position_dodge(0.8))+ xlab(label="Traitement")+ ylab(label="Poids mille graines (g)")+ theme_bw()+theme(panel.grid = element_blank())
Correlation between variables Two-way correlation	library(agricolae); library(correlation); analysis<-correlation(Fert_Tournesol, method="pearson") pairs(Fert_Tournesol)

Principal component analysis
With Factominer
 	
Eigenvalue and cumulative variance
Displaying cos2 and contrib variables
	res.pca1 <- PCA(Fert_Tournesol, scale.unit=TRUE, graph=F) ; plot.PCA(res.pca1,axes=c(1,2,3),choice="var", col.quanti.sup = "red")
	eig.val<-get_eigenvalue(res.pca1)
	fviz_cos2(res.pca1, choice="var", axes=1);
	fviz_cos2(res.pca1, choice="var", axes=1:2);
	fviz_cos2(res.pca1, choice="ind", axes=2);
	fviz_cos2(res.pca1, choice="ind", axes=1:2)

Appendix 3: Some key stages in the trial

Final dissertation for the Diploma of Agricultural **Engineer| Option: Plant Production**

Printed by Books on Demand GmbH, Norderstedt / Germany